Crying Forests

Liu Jinlong

Crying Forests

Political Ecology in the DPRK

中国人民大学出版社
CHINA RENMIN UNIVERSITY PRESS

palgrave
macmillan

Liu Jinlong
School of Agricultural Economics
and Rural Development
Renmin University of China
Beijing, China

ISBN 978-981-16-9727-2 ISBN 978-981-16-9725-8 (eBook)
https://doi.org/10.1007/978-981-16-9725-8

Jointly published with China Renmin University Press
The print edition is not for sale in China (Mainland). Customers from China (Mainland) please order the print book from: China Renmin University Press.

This Palgrave Macmillan imprint is published by the registered company Springer Nature Singapore Pte Ltd.
The registered company address is: 152 Beach Road, #21-01/04 Gateway East, Singapore 189721, Singapore

Acknowledgments

Many have assisted with the production of this book. Sincere thanks to the many friends, colleagues, supporters, and scholars for their help, advice, hospitality, and encouragement—Patrick Durst, Douglas McGuire, Jerker Thunberg, Eva Muller, Youn Yeachang, Lee Changjae, John Parrotta, Liu Yuanchun, Zhu Xinkai, Tang Zhong, Lv Jie, Shi Yan'an, Zheng Shi, Bir C. Mandal, Dominique Reeb, Christophe Besacier, Aggarwal Safia, Jim Carle, Cora van, Oosten, Lee Bianca, Chun BomKwon, Hosang Kang, Sheng Chunhong, Wil De Jong, Simone Aristide-Oke, Ian Armitage, Qu Guilin, Lu De, Zhang Minhui, and Tu Chenyue.

Special appreciation goes to Patrick Durst, Douglas McGuire, Youn Yeachang, Lee Changjae, Christophe Besacier, Lee Bianca, Sheng Chunhong, Simone Aristide-Oke, Zhang Minhui, Tu Chenyuee, and editors for their comments on all or substantial parts of early drafts. I acknowledge and also sincerely thank Simone Aristide-Oke, as the first reader of the final draft for language editing and critiquing. Their comments and works have greatly improved the quality of this publication. All its errors and imperfections are mine, not theirs.

I gratefully acknowledge the generous financial and in-kind support provided by The Research Office of Renmin University of China, the Asian Research Centre of Renmin University China, and Food and Agricultural Organizations. My appreciation extends to Renmin University of China Publishing Co., particularly Madam Liu Yehua for her professional work, and my colleagues, Liu Yuanchun, Zhu Xinkai, Tang Zhong,

Yan Jinming, Shi Yan'an, Lv Jie, Wu Jin, He Xuqiao, Guan Yaran, and Zheng Shi for their generous inspiration, and logistical and administrative support.

I am particularly grateful to the Forest and Landscape Restoration Mechanism of Forestry Policy and Resources Division, Food and Agriculture Organization of the United Nations, Renmin University of China, Seoul National University, Korean Forestry Services, Korean Forestry Research Institute, and many partners in the DPRK for hosting expert meetings and inviting me to attend relevant conferences.

My special thanks go to Michelle Gauthier, who has sadly passed but remains in my memory, as she initially brought me to the DPRK. I extend my gratefulness to many others, especially those who I came across during my time in the DPRK, including the village chairwomen who slept outside with the trees in the winter, elderly women who brought fuelwood home from faraway hills, and the many people I encountered who were out looking for food and collecting fuelwood during the cold winter months.

In 2018, I lost my notebook computer that contained the final edits of this book. For a long time, I considered giving up writing, but encouragement from all those aforementioned have motivated me to see it through. Without their inspiration, rewriting most chapters of this book would not have been possible, especially with my "poor" memory.

Contents

Acronyms

CDM	Clean Development Mechanism
CIFOR	The Center for International Forestry Research
COP	Conference of Parties
DPRK	Democratic People's Republic of Korea
FAO	Food and Agriculture Organization
FLR	Forest Landscape Restoration
GEF	Global Environment Facility
ICRAF	World Agroforestry Centre
IPCC	Intergovernmental Panel on Climate Change
ITTO	International Tropical Timber Organization
IUCN	International Union of Conservation Nature
MDG	Millennium Development Goal
MoF	Ministry of Forestry
MoLEP	Ministry of Land and Environment Protection
REDD	Reducing Emissions from Deforestation and Forest Degradation
SIDA	Swedish International Development Authority
UNDP	United Nations Development Programme
UNEP	United Nations Environment Programme
UNFCCC	United Nation Framework Convention on Climate Change
UNFPA	United Nations Population Fund
UNSC	United Nations Security Council
WWF	World Wildlife Fund

List of Figures

List of Tables

CHAPTER 1

Preface

Part I

Every time I come to the Democratic People's Republic of Korea (DPRK hereafter), it evokes memories of stories from my childhood. During the composting season in early summer, adults would gather together in the morning and tell stories from the night before. After the children were put to bed, the captain, along with the most capable people in the production team, boarded two cement boats and set off quietly into the dark of night. The place they were headed was very remote, somewhere the children had never been and could never go to. The adults said locals there arrest people from other places. The captain took the capable members to "pick up" leaves of the locust tree. They climbed up the tree, used rubber gloves to pull off twigs and leaves, collected and loaded them on the boat, and brought them back to the team.

Before dawn, the branches and leaves of the acacia trees were mixed with the pre-prepared river mud, buried in the pre-prepared compost pit, and cleaned so that even a single leaf could not be seen. The adults said: After some time, these will turn into good fertilizers, which can be used as the base fertilizer for the rice fields. We can count on this to feed you, they said. They often faced danger, as the village where the forest was located had been ordered to defend the property of the socialist state. At night, an ambush was set up. Someone climbed up the tree and shouted out to startle and catch the thief. The adults were so frightened that they

L. Jinlong, *Crying Forests*,
https://doi.org/10.1007/978-981-16-9725-8_1

could only run away and flee wounded back home. We were curious as to how the adults had escaped arrest, the adults said, "We had all become acquainted with each other during the day, and Mr Li's eldest had married into that village. By night everyone pretended not to know each other. If they really wanted to catch them, then there's no stopping them!" The adults worked hard to feed the children.

I majored in forestry at university, and it was then that I found out it was a kind of acacia (Robinia *pseudoacacia* L.), a legume plant. The twigs and leaves were rich in nutrients and made good composting materials. While studying for my PhD, I discovered that the forests beside the river are state owned in name only, but are actually common pool resources. Fortunately, digging up the grass of socialism for non-state purposes was prohibited—let alone digging up the trees. Otherwise, the trees would be robbed and destroyed in broad daylight.

Once I had reached school age, I also took part in activities similar to so-called tree leaf picking. The children did not pick up tree leaves, but collected the broad bean and pea plants owned by neighboring production teams to use as pig fodder for their family farms. At that time, each household raised one or two pigs for the Spring Festival and to support the family. In most families, it was the responsibility of school children to collect various plant leaves for pig fodder during the school period and summer holidays. The collected materials were planted in graves far away from the village. The grown-ups often spoke of ghosts and monsters seen in those places to frighten us. Strangely enough, there were a few bioluminescence sightings in the dark summer nights. There were also more fireflies than usual, that occasionally mixed with large groups of corpse candles—very similar to the scene of ghosts and monsters that the adults mentioned. Stealing forage meant there was more time left to play—us young boys were not afraid of anything. A group of us went to steal broad beans and pea sprouts from our neighbor's team. We had to crawl along the riverside to the destination in the dark of night. The adults from the village next door would come out to catch us, so we hid on the ground and waited for them to come over, and then ran in the opposite direction. We were very proud that we had never been caught. Upon returning to the village after university, the old villagers made fun of our childhood antics, often recalling specific times when we stole bean sprouts or when one of us was so scared that our shoes fell into a ditch when trying to escape. Oh, I finally understand now, it's not that we ran fast, but that they never intended to catch us. These are all things that happened during the People's Commune in my home village.

In the 1990s, a modern version of a high-tech cat and mouse game took place in the forest area owned by the Xiaolongshan State Forest Experimental Bureau in Gansu Province, which interprets the Chinese story of common pool resource management. There is a clear line between the enemy and us, but it is impossible to tell who the enemy is. The 800,000 hectares of Xiaolongshan State Forest Experimental Bureau was on the one side and the community with millions of residents located on the other. Through measures, such as working in the forest farm, making friends, bribery, deterrence, and hiring threats, the local people established alliances and coordinated the division of labor, such as illegal logging, smuggling, processing, and sales, and also purchased vehicles with greater power than patrol jeeps to track the forest patrollers. They tried their best to disintegrate the forest farm's ability to catch loggers. The two sides launched espionage warfare, intelligence stations, psychological warfare, guerrilla warfare, and hand-to-hand warfare, using tactics such as enticing the tiger to leave its mountain lair, beauty traps, money traps, diversion tactics, inflicting injury on oneself to win the enemy's trust, and the empty fort strategy.

Employees of state-owned forest farms can only obtain income and bonuses from timber fines and confiscations if they arrest illegal loggers. The forest farm managers adopted various methods such as strengthening enforcement teams, punishing cover agents and irregular patrols, randomly arranging personnel, and adopting temporary patrolling routes to capture pirates. Leaders of state-owned forest farms met with local leaders and begged locals to cooperate with combating illegal logging. The two parties were enthusiastic and friendly at formal meetings, with strong statements, discussing cooperation in combating illegal logging. Secretly or after official events, local supervisors asked forest farms to open up the net so people could steal wood to buy some "salt."

Illegal logging in Xiaolongshan eventually alarmed Gansu Province. The head of the Provincial Public Security Department was sent down to lead the armed forces to eliminate forest loggers. This action hit the timber thieves and trespassers hard, and there were many casualties, but it mitigated the illegal deforestation epidemic in the Xiaolongshan forest region. During the 1998 China floods, Premier Zhu Rongji promoted the "Natural Forests Resources Protection Program," shutting down timber factories and markets, and completely blocked off illegal logging. Gone were the days of propping up wages with income from timber confiscation at the forest farm. On the surface, the nation and the people said goodbye to the timber theft era! In fact, the story of Xiaolongshan was not over. In the summer of 2005, it was characteristically rainy. During my very last visit to the community central to my doctoral research, there were two young fellows that got carried away by floodwater when out salvaging for wood upstream. They never returned to the community. The forest area of Xiaolongshan is by no means an isolated case. Under special circumstances, uncontrolled logging is common in many countries in Africa, South America, Asia, and Oceania. As long as the social and economic conditions are right, common pool natural resources, just like in the DPRK, will continue to cry!

Part II

On this planet, countries are divided into two categories, the south and the north. The north includes not only the northern part of the northern hemisphere, but also the southern part of the southern hemisphere, such as South Africa, Argentina, Chile, Australia, and New Zealand. Global industry is concentrated in the north, as are the world superpowers and Europe appears to be getting richer. Among southern countries and regions, there are only a few developed economies such as Hong Kong, Singapore, Kuwait, and Jordan. The DPRK is a typical northern country with a long winter and has committed to the process of self-reliant industrialization. However, the DPRK looks particularly different.

1. The DPRK is the poorest country in the North with a population that runs into the tens of millions. If you don't include countries in the former Soviet Union, then China would be classified as the second poorest country pre-2015. In 1990, China's per capita GDP was a quarter of the DPRK's. By 2015, China's GDP was about eight times the DPRK's, and as of 2020, had reached nine times that of the DPRK.
2. The DPRK is the only socialist planned economy that did not fall during the collapse of the Eastern European socialist camp. It has been predicted many times by the United States and many other countries that the DPRK will collapse, but it has not collapsed yet and remains a socialist state.

3. The DPRK was the only country, among the 60 I have visited, that I did not encounter a beggar.
4. The DPRK is a country that has a queuing etiquette similar to Canada.
5. The DPRK insists on 11 years of compulsory education, providing nutritious meals, and free higher education. In terms of social welfare, Canada is not as good as the DPRK.
6. The DPRK insists on free medical care for all. Hospitals are free, and doctors do not accept hongbao (red envelopes that contain money and can be given as a tip in exchange for better medical treatment), as is often seen in China. If we do not take quality into account, then Canada is also inferior in this regard.
7. In Pyongyang, there are rows of residential buildings with no walls, balconies, or anti-theft nets. They are uniform, like a guard of honor. Pyongyang, at least from the surface, is clean and tidy. Among the cities that I have visited, only Washington has comparable cleanliness on the main streets.

Articles 3–7 can only be realized in developed northern countries. In 2015, the per capita GDP of the DPRK was only half that of Kenya and 60% that of the Philippines. This is unbelievable, but a real phenomenon!

Only by acknowledging these facts can we understand why the Chinese men and women football teams were beaten by the DPRK team, and how the DPRK can win gold medals in the Olympics, while India, which has a population 50 times that of the DPRK, cannot.

However, the forests of the DPRK are not so lucky. In the 30 years since 1990, the DPRK has reduced its forests by about 1 million hectares, forest cover has seen an annual reduction of about 2%, and forest stock has halved and has been declining at a rate of about 3% per year. Among countries with a population of over 10 million, the DPRK is the most deforested and degraded country in the world, and the DPRK's forests are far less abundant than Brazil, Papua New Guinea, Congo Basin, and Indonesia. The DPRK is a mountainous country, and forests are critical to the low land security where farmland and urban areas are located.

A comrade from the Ministry of Land and Environmental Protection of the DPRK criticized the work: "They build roads and dams, and yet trees do not survive. In spring, people plant trees in the mountains and plains and in autumn you can see a lot of them, but I bet when I visit again in two years, it will be hard to see any trace of seedlings."

A community (li) where I worked, on the west coast of the DPRK, has built a coastal shelter forest with the support of the FAO. The chief is a strong woman in her 50s. In order to protect this revetment forest, she slept in the forest in a shack with no heating in severe temperatures that reached minus 20 degrees in winter. Her deeds would be seen as particularly moving in any other country, but this sort of behavior is the norm in the DPRK. One day at dusk, officials from the Ministry of Land and Environmental Protection of the DPRK accompanied me to inspect the reservoir near the community. The forest around the reservoir was well protected. When we returned, we came across an elderly woman carrying a basket of small logs and tree branches back to the village. The elderly woman saw us and tried to get rid of us. I prevented the Ministry of Land and Environmental Protection officials from arresting her, even though her actions were illegal. Watching her figure scurry into the distance evoked memories of my childhood. In order to survive the cold northern winters, it is necessary to keep warm.

One day, officials from the Ministry of Land and Environmental Protection of the DPRK led me to visit a red pine forest. This red pine forest was a few hours' drive from the center of Pyongyang. The trees were about 50 years old and more than 20 m high. It was canopy season, but there were few leaves. Most of the bark on the tree trunks were shedding, and they were dark red, with white spots of different sizes hanging drop by drop. The pine trees were like a group of girls crying toward Pyongyang! That pine forest was ravaged by pine caterpillars. Officials from the DPRK Ministry of Land and Environmental Protection asked for countermeasures. I sighed, the larvae can only be eaten by mobilizing the masses to catch them artificially. Early the next morning, there were significantly fewer people on the Daetong River while I was out exercising. Later, I heard that the citizens of Pyongyang had been mobilized to catch insects. Pyongyang does not want the pine trees to cry, it hopes they can become powerful and mighty soldiers to guard the city.

In 2006, I introduced two sets of automatic controlled spray plant nursery systems to the DPRK National Nursery Station. In 2013, I visited a county nursery farm that had fake systems. The leader of the Ministry of Land and Environmental Protection that accompanied me told me that each county nursery station had installed some automatic controlled spray nursery systems produced by the DPRK itself (obviously infringing on China's intellectual property rights). I carefully checked the imitation equipment. The water pressure was unstable and the power supply was not guaranteed. This imitation system was actually worthless. In the county nursery that I visited, the system could only be used to grow willows, poplars, acacias, and other tree species. However, these species could survive even without the use of this system, as long as they were watered on time.

In 2015, the DPRK set out a plan to green 1.5 million hectares of barren hills in 10 years, requiring more than 200 million seedlings each year. It also planned to establish a comprehensive prevention and control system for fire prevention and plant diseases and insect pests, calling on the people to realize "green water and green mountains"—an important part of Kim Jong Il's legacy. It required the party, government, military, and all people to love forests as they love their own eyes and recreate beautiful mountains and rivers. However, I was particularly worried about this sentence: "We must fight resolutely against the bad elements that destroy the forests." Being a companion to the forest does not mean sleeping alongside tigers—the old women doing illegal logging would definitely be afraid.

Part III

A British person working in the DPRK was one of the very few Brits I had come across that didn't have an air of arrogance about him, he asked: "Will the DPRK collapse?" I replied: "It definitely will, but only time will tell when!" The logic was that the DPRK's heavy industrialization strategy was hit by the disintegration of the Eastern European socialist camp and turned to the strategy of Songun Policy (Army first Policy). The DPRK is relatively weaker than the major powers such as the United States, China, Japan, and Russia, but Songun's strategy still frightens South Korea. For more than 20 years, the Songun strategy has had no effect on the four major powers (ranked in terms of overall strength), except for South Korea. The DPRK can only become increasingly irritable and make noises, in an attempt to be taken seriously.

Secondly, the DPRK is a country deeply influenced by Chinese culture. The DPRK has adopted a more radical approach to tradition than China. For the DPRK, that means removing traditional Chinese culture and establishing a new socialist culture. When I was in the DPRK, there were no Chinese characters anywhere to be seen, except for at the Jiuyue (September) Mountain, one of the DPRK's most famous tourist attractions. I didn't see the tomb. Fortunately, the Ching Ming Festival is still held there, and they still carry out tomb sweeping where people go to visit ancestors. What is the new culture? It is difficult to find new connotations other than order, and behind this order is a powerful control system. The British man feared the failure of this control system. Throughout Eastern Europe in the 1990s and the Middle East in the twenty-first century, he witnessed with his own eyes the rape of young girls in broad daylight on the streets, and ordinary people that led a cruel life. A chaotic society without cultural heritage and with cultural constraints is extremely terrifying.

Thirdly, perhaps it is a characteristic of Northeast Asian culture. The population density in this area is extremely high. Therefore, globalization is the only way forward, bringing out the qualities of Northeast Asians with strong endurance and work ethic in exchange for the world's resources. The DPRK is being suffocated by the big powers. Apart from bullying South Korea, there is no way out. The future of the DPRK and the way out of its predicament relies on integrating into the world system, whether it is willing or not.

I watched children in neatly organized groups playing in street corners and parks near Kim II Sung Plaza, and pedestrians in a hurry, looking for happiness or fear or sadness. There was no difference between the views there and the overpass in Beijing, China, and the view at Times Square in New York. The British man said: In the wars that occurred between European countries, including the two world wars, ideological differences occurred between brothers and sisters, and the people were engulfed in ideological thought and killed in darkness. The United States has inherited fine European traditions. As Presidential candidate Clinton Hillary, wife of the former president of the United States, said: "The U.S. Constitution has endowed us Americans with noble citizens and led the people of the world to freedom and democracy. We must undertake such a mission, otherwise, we will lose leadership in some regions, and others may take their place. This is not in the interests of the United States." The happiness and fear of a person and nation are relative to the expectations of each nation and person. The fear the DPRK has for its national survival is the same as the fear American politicians have of America losing its status as world leader, and the fear American people have of life and everything around them. Americans are afraid of guns and terrorists, while North Koreans are afraid of being considered unfaithful to the leader. But to me, the feasting and greenery of Manhattan's Times Square is no different to the silence of Kim II Sung Square.

The British man raised a question to me, "Which one will collapse first out of the United States and the DPRK?" The DPRK, of course, I replied. The United States has no sign of collapsing yet! He explained to us: This world can only be in a better place if both of these two countries collapse! As long as there are Americans, there will be no rule of law and order in this world. I raised a question back: When China becomes the boss, will there be order in the world? Our children are opposing norms, including traditional values of hard work, and rebuilding new ideas. Do these ideas include the rule of law and order?

In the DPRK, the upper class manages everything in this country, everything is in the name of the people. The luxury cruise ship is called the People's Cruise, and the entrance ticket is equivalent to one-twentieth of average monthly salaries. Catering is not something ordinary people can afford, and the yacht attendants are the most refined women. In the DPRK, egoism of the upper-class elites is similar to China. As long as you give cigarettes to customs inspectors, it seems that everything can pass customs. Information and data are often not available, and they are told to keep everything confidential. In fact, if you have money you can get anything you want. Different departments, bureaus, and offices of the same ministries and commissions will not actively communicate or share information. As long as there is no intervention from the higher-level departments, they will never interact with each other. However, as long as there is food and drink, they will always be present. This is no different from the vast majority of countries in the world. World elites have far more opportunities to learn from each other than among ordinary people.

Although the hotel attendants are not the best, they are selected from thousands, most of them are just over 20 years old. I am often the only foreigner in the huge restaurants, which are much larger than ones seen in Europe and North America, and frequently greeted with good morning, good evening, smiling faces, and professional and attentive service. The smiling sometimes gives off a sense of innocence, but more often than not, you can see through it and the insincerity is quite obvious. Just as leaders are talking about great leadership, great party, and great motherland on the face of it, they are actually discussing what they can do with the money, or more directly, how much will be given for their services!

Since the industrial revolution, global forests have been declining. Forests in many parts of the world are crying, and people there are crying too. On this planet, the dragon kings fight while the fish suffer. The order of mankind determines the order of the forest. The tears of the forest resemble the crying of the lost fish or the defeated little dragon king! "The Crying Forest" hopes to let humans, especially those adults, understand that you are fighting, you are capricious, and you are arrogant. Please pay attention to the forest under your feet. Only when the forest cries can human beings laugh; only when the majority of countries cry can a few countries laugh; only when the poor cries, the rich can laugh! If this is the truth, the end of mankind can be expected.

Part IV

This book aims to provide a comprehensive analysis of various social, economic, and political issues that can help us to understand why forests in the DPRK have been crying for decades, and why the DPRK has such severe deforestation. Deforestation and forest degradation issues in the DPRK have been highlighted as an important international political issue that has implications for food security and energy. The DPRK provides a unique case study for the international community that not only illustrates the complexity of forest-related issues, but also shows that we still have a long way to go in overcoming the decline in global forests.

Forests now receive significant consideration in environmental protection and livelihood improvement debates in many developing countries (Sharma, 1992). This is particularly relevant in the DPRK. Deforestation and forest degradation are now global issues, thus causing concern about the effectiveness of current legal, institutional, and policy frameworks that deal with these problems (Liu, 2009). Many international agencies and national forestry bodies have discussed scenarios that may lead to sustainable forest management practices (FAO, 2001; Wiersum & Rostonen, 2005; World Bank, 2001). One widely accepted scenario involves decentralizing forest resource management; however, the DPRK has the most unsuitable circumstances due to its planned economy and top-down decision-making. It is now understood that the forestry issue is complex. It is no longer solely the prerogative of the forestry sector, but involves many political, social, cultural, economic, environmental, and other broad development issues (Zhang et al., 2000; Liu, 2003). International politics, domestic economic development policies, and governance regimes are likely the root cause of deforestation and forest degradation.

This book explains how political, social, and economic factors have intervened with specific natural conditions to cause deforestation and forest degradation in the DPRK. It also examines the efforts taken by

DPRK authorities and international communities through analyzing political, social, and economic factors, as well as forest resources, the history of forest resource management, institutional arrangements, and environmental concerns. The book will: study land use change and forest cover changes and their drivers; map the history of deforestation and afforestation in the DPRK over the course of the century, problematizing it through international politics, domestic development strategies, and governance regimes; clarify the current national efforts required to address deforestation and forest degradation in the DPRK and explain the associated challenges.

As many international forestry-development aid projects have been conducted in the DPRK since 2005, a great number of references are available, various informal sources can be accessed, and many relevant institutions can be visited at the national and county levels. This book is the result of first-hand data obtained from a 10-year study in the DPRK, as well as secondary data.

During the period 2005–2016, I took advantage of the implementation of various international-aid projects concerning forestry development and rural development in the DPRK, and visited the DPRK at least once each year, staying for just over two weeks every time. Naturally, I had plenty of opportunities to visit forest-related institutions, including the Ministry of Land and Environment Protection, Ministry of Timber Production, Academy of Social Sciences, and similar agencies, in addition to a dozen counties and ris (similar to administrative villages in China). As such, I interviewed officials and technicians regarding issues including regulatory systems, administrative systems, reasons for deforestation, planning, and implementation of afforestation, and forest restoration in the DPRK, while at the same time checking what had been implemented at the provincial, county, and ri levels.

For various reasons, it was difficult to collect accurate data in the DPRK. The officials either provided large volumes of confusing and fragmented data or stated that the data required was not available or could not be provided for security reasons. The collected afforestation and forest restoration data was different from the data reported to the international agencies, including CBD and United Nations Convention to Combat Desertification (UNCCD), and data reported to the international agencies was different in various reports. For instance, the total forested area has remained unchanged since 2000, with minor variations in official reports. Data on forest quality from various sources was difficult to collect and often contradictory, including data from official published sources. Nevertheless, upon visiting many sites, I discovered that large areas of "forest" did not actually have trees or only had a few planted trees (mostly covered by crops), but they were officially recognized as forested areas and administrated by agencies under MOLEP.

To obtain reliable data, group discussions were facilitated in 2012 and 2013, which involved officials, professional foresters, and project practitioners from international agencies, including the FAO and the Swiss Development Agency. In these group discussions, concerned influential stakeholders discussed the following topics: big events in the history of forest resource management and utilization, and their impacts on deforestation and afforestation. These discussions have provided a foundation for this study.

In 2013, I was granted a research project to study the drivers of deforestation and afforestation in the DPRK by Renmin University of China. I cooperated with two people from the DPRK, a technician and an official, who stayed in China during the second half of 2014 to work on this project. We collected information on forest-related norms, including land with forests, non-forest land, and stocking forest volume, and discussed how this data could be collected officially. Then, we attempted to estimate the relevant data as best we could, and cross-referenced with official reports in Korean for internal purposes and English versions that had been reported to international agencies.

In the year 2015–2016, visits to the DPRK were used to verify the data used in this research. Semi-structured in-depth interviews were carried out with key informants, including: retired forest scientists, officials from forest-related ministries and counties, ri managers, and professional staff at research and educational institutes. These key informants were usually storytellers and analysts of nature and historical events related to forests in the DPRK.

Group discussions were facilitated in 2016, which involved officials and technicians from various forest-related agencies in the DPRK, and provided a chance to check and reach a consensus regarding the collected data.

The analysis of data from interviews and group discussions, combined with information derived from situational analysis and case studies, was used to explore the research questions and to arrive at specific conclusions. Hence, different bodies of original yet fragmented data were organized into a format relevant to the investigation. Data treatment measures involved comparisons, analogies, induction, deduction, reasoning, and summarization—each designed to identify existing processes and problems for further investigation.

The target audience of this book include: Academics that specialize in forestry governance and global policy; foresters and governmental agencies in the Republic of Korea; International organizations related to the DPRK; environmental NGOs.

First drafted in Pyongyang, July 19, 2016; Rewritten in Beijing Guyue Garden, October 8, 2021.

References

FAO. (2001). *How forests can reduce poverty*. FAO and DFID.

Liu, J. (2003). Support to private and community farm forestry in China. *Unasylva 212, 54* (1), 57–62.

Liu, J. (2009). Reconstructing the history of forestry in Northwestern China 1949–1998. *Global Environment, 4*, 188–219.

Sharma, N. P. (1992). A global perspective on forest policy. In N. P. Sharma (Ed.), *Managing the world forest: Looking for balance between conservation and development* (pp. 17–33). Kendal/Hunt Pub. Co.

Wiersum, K. F., & Ros-tonen, M. (2005). *The role of forests in poverty alleviation: Dealing multiple millennium development goals*. Wageningen University.

World Bank. (2001, July 30). *A revised forest strategy for the World Bank Group*. (Draft). The World Bank www.worldbank.org

Zhang, S., Zhu, C., & Xiao, W. (2000). *Introduction of sustainable forestry management* (pp. 61–63). China's Forestry Publishing House. [in Chinese]

CHAPTER 2

Physical, Economic, and Institutional Setting

The DPRK is unique due to its political, social, and economic systems. It has a territory of 123,138 km^2, with about 25,610,672 residents as of 2018. The DPRK is unique as it is located in Northeast Asia, a relatively small region, but is situated in close proximity to giant global players, namely Russia, China, and Japan. This chapter attempts to draw out key features of DPRK's physical setting, environment, economy, and governance structure to provide a backdrop for the remainder of this book.

Physical Setting

Geography and Climate

The DPRK is situated in the eastern part of the Asian continent, which is located at geographic coordinates of 37°45′–43°00′ northern latitude and 124°10′–130°41′ eastern longitude. The country borders north-eastern China and the far east of Russia, and is surrounded by sea in the east and the west. The average height of the country is 440 m above sea level, occupied mainly with mountains and hills. The DPRK is made up of 83.61% mountains and hills and 16.39% arable/plain areas. Mountains and rivers are basic features of the country, and the word "Mountain-river" is often used in Korean to represent the country's territory. Similar

L. Jinlong, *Crying Forests*,
https://doi.org/10.1007/978-981-16-9725-8_2

characteristics also exist in the Chinese language, as "shanhe"—mountain-river—is used to represent a similar meaning. Mountainous areas contain steep slope gradients that can typically be seen toward the north and eastern sides of the country, while hills with gentle slopes and plain areas are more commonly seen in the western coast and southern part of the country, forming important national production bases used for agriculture.

The country is situated in the temperate zone and has four distinct seasons—spring, summer, autumn, and winter. Influenced by the Asian continent and surrounding oceans, it is hot and humid in summer, and cold and dry in winter. Average yearly temperatures range from 9 to 10°C, with average temperatures in summer (June–August) and winter (November–February) reaching 24°C and –5.5°C, respectively, with slight regional variation. Annual precipitation varies from less than 1,000 mm to over 1,200 mm—much more than global average precipitation which is about 840 mm).

In general, it is arid in spring and rainy in summer, with high quantities of water available during summer months. Precipitation in spring is 15–20% of annual precipitation but 50–60% in summer. Precipitation in autumn and winter is 15–30% and 3–15% of annual precipitation respectively (*Source* DPRK 2nd Communication on Climate Change, 2012).

Precipitation is uneven in terms of region. In recent years, the summer rainy season has been characterized by monsoon climates, with heavy rainfall in the country's central and southern regions. Thus, flood disasters have affected those areas where major agricultural production (rice production) is based. For example, in 2012 Anju, Kaechon and Dokchon counties, located in central DPRK, were affected by heavy downpours that saw 150–300 mm of rainfall within 3–5 hours.

Fauna and Flora

Given its geographical location, the DPRK has rich biodiversity and is home to many unique species. The DPRK can be classified into dominant ecosystem types comprising forest ecosystems, inland freshwater ecosystems, marine and coastal ecosystems, agro-ecosystems, and wetland ecosystems. However, most fauna and flora are found in forests. Forest ecosystems are most important to protecting the diversity of species in the DPRK.

10,012 plant species have been recorded in the DPRK including vascular plants: 207 families, 1,086 genera, 3,623 species; spermatophytes (seed plants): 181 families, 1,017 genera, 3,384 species with 844 varieties; bryophytes: 781 species; lichens: 583 species, fungi: 2,310 species and algae: 2,715 species (*Source* DPRK 4th Report on Biodiversity, 2011). The DPRK recorded 47 endangered or critically endangered species of flowering plants. A further 106 species are considered vulnerable or near threatened.

Lower plants are primary producers in ecosystems and are important for the conservation of terrestrial and marine ecosystem biodiversity. In the DPRK, it is anticipated that more species of lower plants will be found through intensified survey and research.

The DPRK is home to more than 1435 species of vertebrates from 472 genera of 151 families, 3.2% of the world's total. There are 107 species of 69 genera of 28 families of mammals, including 79 species of 48 genera of 20 families of terrestrial ones that are composed of 11 species of Insectivora, 24 of Chiropter, 2 of Lagomorph, 18 of Rodentia, 15 of Carnivora, and 7 of Artiodactyla. The representative mammals include the tiger (Panthera tigris), leopard (Panthera pardus), gray wolf (Cuon alpinus), wolf (Canis lupus), brown bear (Ursus arctos), bear (Selenarctos thibetanus), sable (Martes zibellina), common otter (Lutra lutra), deer (Cervus nippon), musk deer (Moschus moschiferus), water-deer (Hydropotes inermis), roe deer (Capreolus capreolus), and goral (Nemorhaedus goral). Four hundred and twenty species of 190 genera of 61 families of birds have been recorded in the DPRK, including 64 species of residential, 94 species of summer visitors, 106 species of winter visitors, 83 species of passing-migrants, and 73 species of stray birds, of which migratory birds cover the highest proportion. Representative groups of migratory birds include duck-goose, crane, snipe, egret, spoonbill, stock, and swallow. There are a total of 106 species of three groups, namely the duck-goose, crane, and snipe, of which there are many threatened species.

There are 26 species of 17 genera of 11 families of reptiles, which are represented by Eumeces coreensis, Ancistrodon halys, Trionyx sinensis, etc. Amphibians include 17 species of 8 genera of 6 families, which are represented by Bufo bufo, Rana temporaria ornativentris, R. chosenica, and R. coreana. 8,360 species of invertebrates including about 7 600 species of insects have been recorded so far, and its number is anticipated to increase by a further 3/4 times (*Source* DPRK 4th Report on Biodiversity, 2011).

Economic Setting

The economy is highly planned by the government, which manages and allocates food, employment, housing, health care, and education. The country has been excluded from globalization, economic development, international trade, and the investment system.

GDP

The DPRK is socialist planned economy. For a long time, the DPRK has held a competitive position over the Southern part of the Korean Peninsula. The DPRK is listed as having a centrally planned economy, whereas South Korea was listed as having a developing market economy and is only now regarded as having an industrialized and developed economy as of 2020. During the 1970s, average per capita GDP in the DPRK was higher than that of South Korea. In 1989, per capita GDP in the DPRK (approx.1200 US$) was about 4 times that of China (approx. 270 US$).

In the cold war era, the DPRK was a socialist state that had developed its own industry, including iron and steel, chemistry, cements, machinery, and electronics, which have now been integrated into the industry production systems of other socialist countries. In particular, economic development of the DPRK depended on energy supplies from the former Soviet Union, and it enjoyed cheap and reliable energy over decades. 1988 was a turning point in Soviet policy toward North-East Asia. The former Soviet Union terminated its annual delivery of about one million tons of oil to the DPRK, which resulted in severe consequences for the DPRK's economy as well as the livelihood of those living in the northern part of the Korean Peninsula.

Annual data estimates produced by the Bank of Korea since 1991 provide evidence of various internal and external factors that may have drastically reduced the DPRK's GDP in the last two decades (Table 2.1).

Over the past 20 years, frequent natural disasters and persistent economic sanctions imposed from the outside have coincided with the collapse of the east European socialist allies, which has resulted in economic hardship and a decline in overall development.

GDP has declined from 20,875 million US$ in 1990 to 10,528 million US$ in 1996. Over two decades, per capita GDP fluctuated in line with changing weather conditions and the changing political relationship between the DPRK and its four neighboring giants. Over the last decade,

Table 2.1 Gross domestic product (1992–2018)

Year	*1992*	*1994*	*1995*	*1996*	*2000*	*2004*
GDP (million US$)	20,875	15,421	12,802	10,588	10,608	12,859
GDP per capita (US$)	990	722	591	482	464	546
Year	2005	2008	2010	2012	2014	2018
GDP (million US$)	14,561	16,360	13,475	15,510	17,025	17,487
GDP per capita (US$)	615	683	570	643	696	700

the economy has stabilized and resumed a moderate rate of growth in the major sectors. In 2008, GDP was approximately US$16.36 billion, more than 150% of the level of economic output in 1996 (see Table 2.1). Further flooding in 2007 marked another setback for agricultural production from which the economy has recently rebounded. In 2008, the industrial, agricultural, and construction sectors contributed 46%, 15%, and 6%, respectively, to national economic output while other sectors including trade, services, transportation, fishery, and forestry added a further 33%.

To overcome economic hardship, the DPRK, stimulated by the success of economic reform, introduced economic measures in 2002 to promote growth of the markets and entrepreneurship. These measures included price and wage reforms, greater enterprise authority to sell surpluses, the right to sell excess agricultural produce on the market, and the decentralization of decision-making authority to local production units, which is very similar to reform measures implemented in China in the early 1980s. Subsequently, GDP gradually increased from $10.608 billion in 2000 to $16.36 billion in 2007. However, GDP growth remained weak at 3.7% in 2006 and was negative in real terms in 2007 and 2008. Currency reforms launched in December 2009 were aimed at curbing inflation and, together with several related measures, produced unintended results, including high inflation. These sorts of economic reforms were terminated due to political reasons.

The major industrial components of the DPRK's economy include energy production, coal, timber and mining, metal production and manufacturing, railway transportation, as well as the construction material, chemical, and light industries. These industries rely mainly on domestic natural resources. In 1990, the DPRK reached a peak in its industrial sector and produced 14 million tons of cement, approximately 7.2 million tons of steel, and 2.2 million tons of chemical fertilizer. In the 1990s, there were drastic changes in the Soviet Union and Eastern Europe, which caused a deterioration of the DPRK's trade environment, a decline in foreign exchange reserves, and a sharp fall in the import of raw materials, fuels, and crude oil. All aspects of the industrial sector have faced severe recession. Between 1990 and 1998, coal production decreased from 33.15 million tons to 18.6 million tons; power generation decreased from 22.7 billion kilowatts to 17 billion kilowatts; steel production decreased from 3.36 million tons to 945,000 tons; crude oil imports decreased from 2.52 million tons to 500,000 tons. Due to serious shortages of raw materials, fuels, and electricity, most large and medium-sized enterprises were in a state of suspension or semi-stop production, and the operation rate of enterprise equipment was only 20%. In 2000, chemical fertilizer production decreased to almost nothing, and cement production decreased to below 2 million tons. After 2000, major industrial productions have slightly increased.

Tackling the energy and food production shortages is a main priority, as they are the two major challenges facing the DPRK economy. Projects that involve the construction of small, medium, and large hydro-power plants have contributed to an increase in the rate of hydro-power to coal-power generation, from 5:5 in the 1980s to nearly 6:4 in the 2000s.

Agriculture

The northern and eastern regions of the DPRK are particularly mountainous, accounting for approximately 75% of the total territory. The mountains in the northern and eastern sections of the Korean peninsula extend gently, forming hilly and plain areas in the southwest. The largest hilly areas can be found in the Luling Hills and the Cheling Hills, with many plains interspersed in between. These plains are: the Longcheon Plain in the lower reaches of the Yalu River, the Bocheon Plain and the Anju Plain in the lower reaches of the Cheongcheon River, the Pyongyang Plain in the lower reaches of the Taedong River, the Jaenin Plain in the Jae Ning River Basin, the Yanbai Plain in the Lower Yeecheng River, and the Han River formed by the alluvial Han River Plain, Naepo Plain in the lower reaches of the Jinjiang River, Hunan Plain in the Jinjiang River, Wancheong River and Dongjin River Irrigation, and Gimhae Plain in the Nakdong River Basin. Plain areas in the DPRK have common characteristics, including convenient transportation, dense populations, fertile farmland, and famous orchards, known as the "barn zone" of the peninsula.

Farming requires land. The DPRK has taken active measures to improve food security by supporting large-scale land realignment, constructing artificial reservoirs and waterways, introducing new agricultural technologies, and adopting a two-crop per year farming cycle where possible, which demonstrated success before 1990 and failure thereafter. One achievement was the reclamation of tidal areas. On the west coast, the Taegyedo Tideland reclamation project significantly increased agricultural land in the area. Construction projects of naturally flowing waterways, such as the Kaechon-Lake Taesong waterway and the Paekma-Cholsan waterway, have been used to irrigate lowland arable lands in western and southern areas, which were previously irrigated using pump stations and power supplies. Construction of large-scale standardized farmland has been implemented across the country and carefully divided slopes merged on gentle hillslide slopes, both of which have been conducive to reducing labor and costs, and improving the conditions of agricultural machinery usage. This fragmented land has expanded and turned into large-scale standardized fields. Throughout the phase-wise activities implemented from 1996/1997 to the early 2000s, most of these small-patched arable lands had been turned from medium to large sized standardized crop fields.

Modernized farming required fertilizer as well. Until the 1980s, the country supplied sufficient chemical fertilizers to the agricultural sector and facilitated intensive farming with high, sometimes even the highest, crop yields. However, during the last three decades, domestic chemical fertilizer production had dropped as it could not meet rural demands. Economic sanctions interfered with the import of fertilizers, which has had adverse impacts on agricultural production. To cope with this situation, the DPRK has chosen a method to renovate and modernize its domestic chemical production plants (such as the Anju chemical factory, and the Hungnam chemical fertilizer factory) to meet chemical demands using its own products. To do this, the DPRK invested in the renovation and modernization processes of these plants, but could not produce sufficient chemical fertilizers to meet demand.

In the DPRK, farming is heavily dependent on the country's irrigation system. Take the Taedong River as an example, the Taedong River has a total length of 439 km and a drainage area of 20,300 square kilometers.

There are hundreds of tributaries in the basin and, due to an abundance of rainfall and mountains, most of them are mountain streams. The upstream flows through lofty mountains, with a fast flow rate and abundant water resources in the rainy season. The middle and lower reaches are gentle and easily irrigate farmland. The rainy season in the DPRK is concentrated at the turn of summer and autumn. Due to the destruction of vegetation in the upper reaches of the Taedong River, floods often occurred and attacked low reaches of the river and inundated the farmland. The West Sea sluice, one of the five sluices constructed on the Taedong River, achieved comprehensive benefits such as flood control, irrigation, shipping, water supply, and power generation, and can irrigate around 20,000 ha of farmland.

The DPRK paid great attention to farming as part of a national strategy focused on self-sufficiency. Rice and maize are the most important crops, followed by wheat, barley, sorghum, beans, and potatoes. The status of main cultivation land use can be seen in Table 2.2. The area of land cultivated for rice production has remained largely stable since 1996, while maize cultivation has declined and wheat and barley cultivation has increased (see Table 2.2). After 2010, there was a gradual increase in land cultivated for potatoes, as the DPRK set potatoes as a key part of its national strategy to overcome the food shortage.

Statistical records of the DPRK's grain output began in 1961, when it was 3.58 million tons, and has hovered between 3 and 4 million tons thereafter. From 1969 to the early 1970s, it exceeded 4 million tons. In the 1980s, it rose again to between 5 and 6 million tons. In the early 1990s, it broke 9 million tons and reached a record high of 9.31 million tons in 1992.

Table 2.2 Areas of seeding and cultivated land of main crops (Unit: 10^3 hm^2)

		1996	*1998*	*2000*	*2002*	*2004*	*2008*	*2011*	*2014*
Seeding areas (thousand ha)	Rice	575	574	531	578	–	565	571	525
	Maize	584	588	492	493	–	499	503	532
	Wheat, Barley	64	79	122	126	–	108	90	54
Cultivated land area (thousand ha)	Orchard	159	159	157	158	143	–	–	–
	Mulberry	–	92	92	89	84	–	–	–

Source Environment and Climate Change Outlook, 2010; http://m.haiwainet.cn/middle/232591/2015/0619/content_28851600_1.html

Natural disasters occurred frequently in the DPRK in the 1990s. Natural disasters such as droughts, floods, tidal surges, hailstorms, typhoons, and extreme winters had a severe impact on agriculture to varying degrees, which in turn had a great impact on agricultural production, especially in the mid-1990s. The freezing disaster in 1993 and the hail disaster in 1994 caused grain output to drop sharply from 6.6 million tons in 1990 to less than 4 million tons in 1995. Another flood occurred in 1996, and grain output was only 2.5 million tons. There was a drought in 1997, and grain output was 2.7 million tons. In 1998, there was a tsunami and frost. During this period, the DPRK's grain output dropped to its lowest ever. In 1999, grain output was the highest since 1994, and it was only 4.22 million tons. In comparison with the minimum food requirement of 5.18 million tons that was officially announced by the DPRK, the food shortage gap was 960,000 tons. Food imports of 1.25 million tons were continually needed, and one-third of the population faced food shortages with the World Food Programme only able to provide for 1.4 million of the 6 million people who needed food aid (UNDP, 2011).

In 2010, the DPRK outputted 4.52 million tons of grains such as rice and corn, which was similar to levels seen in the 1970s. In 2011, the DPRK's rice planting area covered 571,000 hm^2, corn planting 503,000 hm^2, potato planting 34,000 hm^2, and soybean planting 131,000 hm^2. In addition, there were 400,000 to 450,000 hectares of land cultivated by state or line agencies. The arable land was mainly located in the central and western coastal areas. In 2010, the DPRK's grain output hovered between 3.5 million and 4.5 million tons, and there was a shortage gap of at least 1 million tons per year. The global hunger index score for the DPRK was 28.8 in 2015, an increase of 12.5 from 2014, and classified as "serious" (IFPRI, 2015).

In 2014, the total farming area amounted to 1.4291 million hectares, with an estimated output equivalent to 5.713 million tons of grain products, an increase of 48,700 tons on the previous year. Among them, the rice planting area was 525,000 hectares and the expected output 2.626 million tons; the corn planting area 532,000 hectares and the expected output 2.349 million tons. The DPRK's food production in 2018 was at its lowest level since 2008 at 4.951 million tons, a decrease of 503,000 tons from 2017. In 2020, total grain output was 4.4 million tons, a decrease of 5.2% from the previous year. The agricultural land was about 1.85 million hm^2, of which the cereal planting area was 1.4 million hm^2.

Energy Sector

Construction projects of large-scale hydro-power plants: Anbyon, Taechon, and Huichon hydro-power plants were used during periods of hardship. Such construction projects contributed to an increase in the rate of hydro-power generation to coal-fueled generation, from 5:5, to around 6:4. Promotion of small and medium-sized hydro-power plants: the DPRK pursued a policy of constructing small and medium-sized hydro-power plant/stations in rural areas to meet their own energy demand and, as a result, many plants and stations were built and are still under construction, such as the Kumyagang, Wonsan, and Ryesonggang hydro-power plants. Construction of hydro-power plants in this way has led to an increase in waterbody areas and a decrease in other landforms, namely arable lands.

Industry

In 1954, the DPRK introduced the following industry policy: "prioritising heavy industry development while also developing light industry and agriculture," which led to a period of rapid individualization. From 1954 to 1963, after the Korean War, the growth rate of industrial production reached a terrifying 34.8%. By the 1970s, industrial output value accounted for 75% of the country's GDP. In the late 1960s, electrification was popularized throughout the country. Every rural household was connected to electricity, and electricity was generally used as a source of mechanical power and heat. The per capita electricity consumption reached 1,184 kilowatt-hours, which was five times the average electricity consumption of China at the time. China's electricity consumption per capita did not reach this level until 2004. As a result of industrialization, the DPRK achieved mechanization of agriculture in the late 1970s. 95% of transplanting seedlings, 80% of harvesting, and the entire agricultural irrigation all used agricultural machinery. In 1970, the DPRK officially declared that it had become a socialist industrial country with modern industry and developed agriculture.

In the 10 years since the end of the Korean War, the DPRK's average annual economic growth rate reached an astonishing 25%, ranking first in the world—a figure that stunned the world. There were only two industrial countries in East Asia in the 1960s and 1970s, one was Japan and the other, surprisingly, was the DPRK, known collectively as the "East Asian

Twins." The DPRK was recognized by the Eastern European socialist camp as "a miracle of the Far East" in terms of economic development and became the target of learning and catching up at that time.

However, everything in the DPRK, including industrialization and agricultural mechanization, requires sufficient fossil energy to drive economic development and social affairs. The domestic industrial system and agricultural machinery are all operated by petrol and diesel. The DPRK has no crude oil storage and uses to buy oil at a "friendly price" from the former Soviet Union and other countries in The Council for Mutual Economic Assistance (Comecon), or through bartering for oil with Comecon. When the former Soviet Union disintegrated in 1991, the glorious environment that the DPRK relied on completely disappeared. After the disintegration of the former Soviet Union, it no longer sold oil to the DPRK at low prices, but required the DPRK to use hard currencies such as US dollars or gold to buy at market prices.

The golden period of the DPRK relied on help in the form of orders from Comecon countries. Foreign trade was carried out within Comecon, but there was little trade with other countries. During that period, the DPRK didn't have sufficient gold or US dollars, but it was not necessary to have hard currencies. Before the collapse of the Soviet Union, trade between the Soviet Union and Eastern European countries accounted for more than 80% of the DPRK's total trade, and trade between the Soviet Union and the DPRK accounted for only about 53%. However, after the disintegration of the Soviet Union up until 1994, Russia's trade with the DPRK dropped sharply from 53 to 8%, and in 1995 it dropped further, to 3%.

The DPRK's crude oil imports, its lifeblood at the time, dropped sharply from 18.47 million barrels in 1990 to 3.88 million barrels in 2014. Without oil, it was impossible to operate power plants, factory equipment, and industrial machinery, and the upstream and downstream industrial chains were completely broken. Highly mechanized agriculture was reverted back to the farming technology used in the peasant era. The modern industrial system had effectively been completely shut down.

The DPRK is a socialist industrial country and proud of this title even in the present day. The DPRK has strong industrial capacity, including high-tech military industry that can be exported to other countries for foreign currencies. The four global giants have been engaging in an attempt to stop nuclear activity in the DPRK. After the USSR collapsed, the DPRK aimed to develop a self-reliant national economy, which has been the case to some extent since 2000. In the last 10 years, the DPRK has developed its light industry which gradually helped to improve people's livelihood. In the DPRK, key industries include the electric power industry, coal industry, metalworking industry, mining industry, engineering industry, construction industry, chemical industry, and the light industry.

Population

After the Korean War, the DPRK's population grew continuously, even throughout the great famine of the 1990s, but population growth slowed down during this time. In 2008, the population of the DPRK was approximately 24.05 million and had an annual population growth rate of 13.4‰ (13.5‰ and 13.3‰ for males and females, respectively). The annual population growth between 1991 and 2008 was 0.86‰. The total population in 2019 was 25.53 million, an increase of 0.5% from the 24.93 million recorded in the 2015 survey. The average life expectancy of local people is 71 years old in the DPRK, which is close to the world average of 72.6 years. Population increased by 4.77% in 10 years between 2011 and 2020 (see Fig. 2.1).

In 2008, about 60.6% of people lived in urban areas and 39.4% in rural areas (UNFPA, 2008, 2014). The total number of households was 5.887 million, a 22.6% increase on the numbers recorded in 1993. The United Nations Human Settlements Program (UN-Habitat) released the "Report of World's Cities" in 2020. There were 16.12 million people

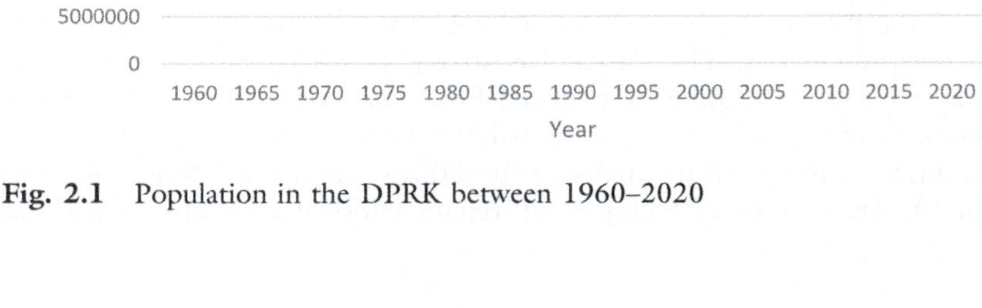

Fig. 2.1 Population in the DPRK between 1960–2020

living in cities in the DPRK, with an urbanization rate of 62.4%. The DPRK's urbanization level is higher than the world average of 56.2% and the Asian average of 51.1%. The report predicts that the urban population of the DPRK will increase to 16.816 million by 2025 and 17.531 million by 2030, accounting for 63.8% and 65.6% of the total population, respectively. The level of urbanization in China will increase to 66.5% in 2025 and 70.6% in 2030, and it will surpass the DPRK in five years. Cities in the DPRK, in particular Pyongyang, left a deep impression on me due to its cleanliness, spacious housing, and good medical care.[1]

Government

The DPRK is a socialist country, governed by the Supreme People's Assembly and the Cabinet. As the highest organization of state power, the Supreme People's Assembly adopts national strategies and legislation, and supervises enforcement. The Cabinet is a central government organization that leads and manages administrative affairs and is responsible for economic development. The Cabinet supervises various subordinated ministries, including the Ministry of Land and Environment Protection, the Ministry of Timber Production, and the Ministry of Agriculture. The Ministry of Timber Production manages most state forests and administrates timber products, including timber harvest planning, harvesting, collection, distribution, and processing of timber products.

The DPRK is divided into five zones (north western, north eastern, central, south eastern, and south western) based on geography and administrative areas. There are nine provinces: Ryanggang, North Hamgyong, South Hamgyong, Gangwon, Jagang, North Pyongan, South Pyongan, North Hwanghae, and South Hwanghae. Provinces are further divided into cities (districts) and counties with a total number of 210. The ris are the lowest administrative unit within the county (gu- and dong-) and one county has approximately 15–20 ris. Counties consist of ris, the smaller geographical units, and up, the county seat. Cities (districts) consist of dongs. In large cities, dongs are grouped into districts.

[1] UN-habitat. 2020. Report of World Cities: Value of Sustainable Cities.

References

DPRK. (2011). *4th report on biodiversity* (English).

DPRK. (2012). *2nd Communication on climate change.*

International Food Policy Research Institute (IFPRI). (2015). Global hunger Index—Armed conflict and the challenge of hunger.

UNDP. (2011). *Country programme for the democratic People's Republic of Korea (2011–2015).*

Unite Nations Population Fund (UNFPA). (2008). *DPRK population census national report.*

United Nations Population Fund (UNFPA). (2014). DPRK socio-economic, demographic and health survey; DPRK Central Bureau of Statistics. (2008). DPRK 2008 population census—National report.

CHAPTER 3

Major Environmental Issues

Environmental concerns in the DPRK are related to population growth and socioeconomic development, as is the case in many other developing countries. Environmental concerns in the country have, to some extent, been triggered by the over-exploitation of natural resources and environmental pollution due to farmland expansion and industry emissions. However, the DPRK is a special case, and its environmental issues are largely affected by international politics. International sanctions have severely weakened the DPRK's economic development capabilities, and so there has been an increase in over-exploitation of natural resources in an attempt to promote economic development and in turn improve people's livelihood, which has had a negative impact on the ecological system and caused environmental damage.

Flood and Natural Disasters

In the DPRK, natural disasters are closely related to the country's natural environmental conditions and the impact of global climate change. The disasters are caused, in many cases, by floods, droughts, and tidal waves. Natural disasters have increased in intensity and severeness due to accelerated development activities such as deforestation, and human induced climate change.

L. Jinlong, *Crying Forests*,
https://doi.org/10.1007/978-981-16-9725-8_3

This represents one of the fastest rates of warming among Asian countries. The DPRK has been seriously affected by climate change. In the period 1918–2000, average temperatures increased by 1.9 °C, more than double the global average of 0.7° C. This temperature rise has changed rainfall patterns and has increased the incidence of extreme weather occurrences like floods and droughts, which in turn has impacted socioeconomic development and people's livelihood.

Over the past three decades, the frequency of flooding events has increased and caused tremendous damage. On October 10, 2001, 347 mm of rain fell in the Wonsan area in just 12 hours. In July of 2006, 448 mm of rain fell over three days in the Yangdok area (*Source* DPRK, 2012). In the summer of 2007, large areas of the north and south regions of Pyongan, Hwanghae, Kangwon, and Hamgyong Provinces suffered from torrential rain over three days which resulted in widespread damage to arable land and infrastructure, and destroyed a large portion of agricultural production. In 2021, between 18:00 August 1 and 19:00 August 2, average rainfall in South Hamgyong Province reached 113 mm, and the rainfall in some areas exceeded 300 mm. In many areas of South Hamgyong Province, heavy rainfall triggered floods, causing rivers to rise and break through dams. More than 1,170 houses were destroyed or submerged, and more than 5,000 local residents were urgently moved according to a DPRK official report. Large areas of local farmland were also submerged, and many bridges, roads over 16,000 m, and river dams over 8,000 m were destroyed by floods.

Heavy rainfall in deforested areas caused landslides that deposited large amounts of soil and sand, land cover, and fallen trees and forests in the valleys, and raised the riverbed level causing subsequent flooding to occur more easily. In Yangdok County, there were several landslides in 2006 and 2007 that damaged railways and buried houses. In 2007, rivers had reduced capacity from previous landslides, which made it difficult to move large volumes of water due to high sediment loads. Thousands of hectares of paddy fields had flooded.

Droughts are severe too. Throughout May 2012, there was almost no rain in the west coast of the DPRK, and there was an unprecedented drought. The amount of precipitation was the lowest since 1962. The soil moisture was about 55%, which seriously affected spring sowing and transplanting.

In 2015, the DPRK reportedly suffered the worst drought in a century. According to official figures from the DPRK, average rainfall from January to May 2015 was 135.4 mm. Although slightly higher than the same period in the previous year, it was only about 74.2% of annual rainfall, and abnormally hot for late May, which contributed to extreme droughts. Due to droughts and little rain, rivers in many places dried up, and reservoir water levels dropped to the lowest level in history. As of June 8, a total of 441.56 million hectares of rice fields in the DPRK have been transplanted, but at least 136,200 hectares have dried out due to droughts. The North Yellow Sea, South Yellow Sea, South Ping An, and South Hamgyong Roads, which are regarded as the DPRK's granaries, were the hardest hit. According to reports, droughts have destroyed nearly 80% of rice seedlings and 58% of rice fields in the North Yellow Sea and South Yellow Sea.

The DPRK's national average rainfall in the first five months of 2019 amounted to 54.4 mm, the lowest level since 1982. There was little rain in the north, and there was no rainfall at all in the central and southern regions such as Pyongyang, Nampo, Huanghaidao, and Gangwon Province.

Previously, droughts commonly occurred from March to May, but in recent years dry periods have increased and continued right through to July and August. 1990, 1995, and 1996 were years of abnormally heavy rain, whereas the country witnessed a spell of dry weather in 1997 and 2000.

Land Degradation and Soil Erosion

Land degradation is also one of the main environmental concerns in the country. It is closely related to the specific natural conditions of mountainous areas, changes in the climate and population growth. Soil erosion is the main cause of soil degradation. Soil erosion is more severe in sloping lands with a gradient of over 10°, which occupy over 40% of non-paddy field areas. The loss of soil by erosion on sloping lands is about 40 to 60 tons/ha per year. In severe cases, it exceeds 100 tons/ha per year (*Source* DPRK, 2012). During the last 25 years, on average 47.8–50.8 mm has been lost from the top soil layer—about 2 mm per annum, which has had adverse impacts on agricultural production with a decrease of 23–25% during the same period. As a result, sloping land with less than 30 cm of top soil layers increased by 25%, and those with over 50 cm of soil layers decreased by 30% in the 25 years before 2010. Furthermore, areas of low humus content are increasing every year because of diminution of basic fertility index and humus of soil. During the 17 years before 2010, low humus areas increased by 8.9%, and high humus areas decreased by 5.5%.

Erosion in forest land has led to a decrease in organic material content and forest water storage capacity, and also an increase in landslides caused by floods. In particular, due to a lack of food, cultivation of forest land has led to further erosion. The rate of landslide occurrence at each gradient can be seen in Table 3.1.

Land degradation is a critical issue due to the soil erosion which occurs directly as a result of deforestation and unstable crop cultivation in sloping lands. Unlike other environmental issues, land degradation is not easy to recover once it has occurred. Currently, soil erosion in sloping lands is about 40–60 tons/ha/year with a maximum value of over 100 tons/ha/year. Assuming that the formation of 1 mm layer of soil takes 25 years, the restoration of 50 mm layers of soil lost within 25 years through erosion would take 1250 years. In addition, land degradation occurs in rice fields converted from tidelands, through water-logging, loss of agricultural lands due to infrastructure construction, intensive mono-cropping, over-grazing, and damage of coastal areas by tidal waves. They all diminish the productivity and regeneration capacities of ecosystems, leading to insecurity of food production and an increased vulnerability to severe and frequent disasters.

The forestlands are also destructed and degraded due to the repeated natural disasters and economic hardship of the 1990s. Of a total of 9 million ha of forestlands, about 1.5 million ha have been deforested and converted to non-forested lands, or slashed and burned to cultivation areas. On the other hand, over-logging in large areas of industrial forests (timber production forests) has reduced the forest products including standing timber volume. Deforestation and forest degradation have led

Table 3.1 Rates of landslide occurrence at different gradients

Gradient (unit: °)	*20 <*	*20–25*	*25–30*	*30–40*	*>40*
Rates of landslide occurrence (unit: %)	0–0.5	3.5	14.3	77.3	4.9

to the acceleration of biodiversity loss, soil erosion, and deterioration of hydrological dynamics as well as more frequent natural disasters.

Loss of Biological Diversity

The DPRK is very important to conservation of biodiversity in the temperate zone, as it has great biodiversity, despite its small territory, due to its geographical position that connects the continent with the ocean. The biodiversity loss at ecosystem level appears mainly in the form of deforestation, i.e., the destruction and fragmentation of habitats, and the degradation of marine, coastal, freshwater, and wetland ecosystems.

At species level, 10 critically endangered species, 42 threaded species, 76 rare species, and 26 endemic species have been identified for higher plants, which make up 4% of the total higher plant species. Similarly, 9 critically endangered species, 29 threatened species, and 119 rare species of vertebrates have been recorded, equaling about 11% of their total. In fact, a number of species including tiger, musk deer, medicinal plants, and marine aquatic resources face threat of extinction due to over-harvest for medicinal and economic purposes.

The CBD 2010 biodiversity target was adopted in the Decision (VI/26) of COP that emphasized the importance of identifying status of selected ecosystems and habitats, changes in status of threatened species, and arrangement of selected species. The threatened plant species (seed plant) accounts for about 4.54% of the total number of species.

Threatened plants are classified, according to the criteria outlined by IUCN, into extinct (EX), critically endangered (CR), endangered (EN), vulnerable (VU), and near threatened (NT) species as shown in Table 3.2.

Among these 153 threatened species, there have been many endemic cases, including Picea tonaiensis (gymnosperm), Celtis cordifolia, Rosa jaluana, Nymphaea minima, Prunus choreiana, Pedicularis ishidayana,

Table 3.2 Threatened Species of Plants in DPRK

Taxon	*CR*	*EN*	*VU*	*NT*	*Total*
Gymnosperm	1	2	3	3	9
Angiosperm	15	29	48	52	144
Total	16	31	51	55	153

Source DPRK 4th Report on Biodiversity (2011)

Saussurea myokoensis, Euphorbia hakutosanensis, and Pentactina rupicola.

The main threats to plant species are: 1) environmental changes in habitat; 2) over extraction (Panax ginseng, Astragalus setsureianus, Asarum heterotropoides var. seoulense, Thymus quinquecostatus, Dioscorea nipponica, Scopolia parvipolia, and Gastrodia elata); and 3) deterioration of the eco-environment due to decreased forest land.

Although there are about 200 protected areas and protected forests of various kinds throughout the country, they are managed poorly due to a lack of proper management capacities. Besides, the unsustainable use of biodiversity components in other areas is also a major challenge to the conservation of biodiversity.

Degradation of Water Quality

The DPRK has an abundance of water resources. However, they are depleting at an increasing rate for various reasons including population growth, urbanization, industrial and infrastructure development, and deforestation.

Water quality deteriorates through increased solid waste sedimentation and eutrophication, which has negative impacts on the supply of drinking and industrial waters as well as on biodiversity. Water quality varies by location and also by season. The majority of water pollutants comes from the discharge of industrial wastewater and untreated sewage, particularly in rural areas where facilities are inadequate. Runoff from agricultural land is another source of contaminants.

Degradation of Marine and Coastal Ecosystems

The DPRK has unique marine and coastal ecosystems, as it is surrounded by seas on both the eastern and western sides. However, alongside the increasing impacts of climate change, there are increasing development and exploitation activities on marine and coastal ecosystems, which are associated with high population density and limited natural resource availability. Many coastal areas including wetlands are largely exploited through human development activities including salt farming, overharvesting of shellfish, and the reclamation of tidelands for conversion to cultivation areas. They have inevitably led to the rapid alteration of coastal wetland ecosystems and a reduction of coastal aquatic resources.

Air Pollution and Solid Waste

Air pollution is a major environmental concern in the DPRK. Due to old-fashioned and inefficient emission control facilities in industrial processes, air pollutants including sulfurous acid gas and aerosols are emitted that exceed national standards, causing air pollution in urban areas. The low combustion efficiency of fossils fuels, mainly coal, and the decreased energy conversion efficiency also contribute to increased carbon dioxide emissions. The deforestation and land use change are additional factors that reduce carbon sinks and contribute to global warming. On the other hand, yellow sand storms that occur more frequently in Asia have brought increasingly severe damages to the environment and human life.

Solid waste is a function of production activities and life process of human society. Currently, the lack of sound management system of solid wastes in urban and rural areas has negative implications on the hygiene and sanitary environment, as well as the soil, water, and air quality.

References

DPRK. (2011). *4th report on biodiversity* (English).

DPRK (Democratic People's Republic of Korea). (2012). *Environment and climate change outlook.*

CHAPTER 4

Land Use and Land Use Change

Land Use Category

The total land area of the country accounts for 123,138 km^2 and is made up mainly of mountainous forest land, with fewer areas of arable land. DPRK land can be divided into six categories—agricultural land, forest land, settlement areas, industrial land, water area, and other land. In 2010, forest cover was approximately 92,062 km^2, agricultural land 18,680 km^2, industrial land 1,844 km^2, water bodies 7,683 km^2, residential land 1,595 km^2, and the rest for miscellaneous use (see Table 4.1).

Most arable land is located downstream of major river courses in the seacoast regions. It is small relative to the population size (0.08 ha/capita) and lower than the global average. Intensive farming has begun in these limited areas to keep up with rising food demand in the country.

The largest proportion is the forest land which makes up 72.5% of the total land area. Forest land and its natural resources are vital for national economic development and high quality of life, especially in rural areas. This does not mean that all forest land is covered by forests. The land was officially categorized as forest land and was previously designated to forestry. However, in the last 20 years, many forest areas have degraded, with a large proportion deluded or barren and yet remain part of the forest land category.

L. Jinlong, *Crying Forests*,
https://doi.org/10.1007/978-981-16-9725-8_4

Table 4.1 Officially recognized Land use in 2010 (km^2)

Land use category	*Forest land*	*Agricultural land*	*Industrial land*	*Water bodies*	*Residential land*	*Miscellaneous use land*	*Total*
Area	92,062	18,680	1,844	7,683	1,595	1324	123,188

Source DPRK 2nd Communication on Climate Change (2012)

Residential areas cover 1,659 km^2, 1.3% of the total DPRK territory, including dwelling houses and public establishments such as houses, roads, factories, and park covers.

The land area of lakeshores, reservoirs, and river flow stations combined is 7270 square kilometers, accounting for 6.0% of the total land area. The DPRK has about 100 natural lakes, 1,700 artificial reservoirs, 38 major rivers, and 10,208 streams. The total length of all streams is approximately 64,855 km, of which major rivers account for 5,355 km and medium-sized streams approximately 59,500 km. The DPRK has a high density of watercourses (0.4–0.6 km/km^2) in comparison with most countries. Water areas cover 6.0% of the country and hydropower plants have caused an increase in total water area.

Large-scale industrial enterprises, mines, coal mines, and other industrial lands total 2020 square kilometers, 1.7% of the total territory. Other land covers 640 square kilometers, accounting for 0.5% of the total land area.

Natural parks and conservation areas: Protected areas of various types, including historical cultural relics, archaeological sites, scenic spots, and natural monuments, cover 879,275 ha in total.

Land Tenure Arrangement

Land in the DPRK belongs to the state or cooperatives. Land belongs to the people and the public, and individuals are unable to buy, sell or take land for personal ownership. Land under state ownership is owned by the people. Land owned by cooperatives is collectively owned by laborers participating in the collective economy. Based on the development of the agricultural economic system and the voluntary participation of all members of the cooperative, cooperative land can be confiscated and given to the people. Cooperative farms, institutions, enterprises, organizations, and citizens can use state land in many ways for the benefit and happiness of the people. Co-operative farm staff are permitted to have 20–30 level ground next to their house[1] according to the co-operative farm regulations.

Land management policies and legislation are executed by land use organizations and control organizations of land management. The Ministry of Land and Environment Protection (MoLEP) is responsible for the supervision of land in the country. MoLEP has institutional structures including departments for Forest Management, Rivers, and Streams Management, Road Construction and Management, and Land Supervision so as to supervise and coordinate national land use. There is a land and environment protection bureau set up in each province and a department of land and environment protection set up in each city and county. The Department of Forest Management under MoLEP conducts management and supervision of forest land. The Ministry of Forestry produces, creates, and manages forest products and the cyclic felling areas are selected by MoLEP. The Ministry of Agriculture manages agricultural land directly.

Each province has an agricultural management committee, each city (county) has a co-operative farm management committee and each ri has co-operative farms, and so agricultural land is managed and used collectively. Farmer households have the right to use up to around 270 m^2 of home garden land and associated products.

However, the management and/or usage rights of the land is differentiated in various ways. For example, cooperative farms and rural factories situated close to forests (designated state areas) can be used, after obtaining permission, for firewood plantation or other purposes. Cooperative farmers can conduct illegal farming in degraded state forest areas, and since 2008, around 30,000 ha of state forest area has been designated to farmer cooperatives for agroforestry.

Land Use Change

There is no readily available statistical data on land use change, but it is understood that there have been no significant land use changes over the last few decades in the country. As seen in Table 4.2, agricultural and

[1] Level ground is the area unit of the Japanese traditional measurement system, 1 level ground = 3.3057 m^2.

Table 4.2 Land use change from 1990 to 2011 (km^2)

Index	*1990*	*1993*	*1996*	*2002*	*2011*
Forest land	89,455	8,8235	88,324	88,285	87,170
Agricultural land	20,212	20,698	20,866	20,856	18,680
Industrial land	1,874	1,944	1,974	2,003	1,844
Water bodies	7,041	7,141	7,210	7,210	7,683
Residential land	1,369	1,507	1,557	1,597	1,595

Source DPRK 2nd Communication on Climate Change (2012)

forest land has decreased gradually overtime, while residential and public land, such as houses, roads, factories, and parks, have increased.

Land reform was implemented in 1946. Soon after liberation, the DPRK announced the land reform legal declaration on March 5, 1946, which confiscated land from landowners that owned over 5 ha of land, and distributed it to tenant farmers and farmers who used to own the land. This introduction of individual ownership of land boosted agricultural production.

After the Korean War, the DPRK initiated socialist reform to transform land ownership from individual to cooperative ownership. This policy implementation was completed by 1958, by which time all arable land had been turned from individually owned to cooperative and state-owned property. Not long after, policy was introduced to reclaim new arable land from tidelands, hillsides, and terracing fields, which spurred an increase in availability of arable land.

Land is managed according to the land categories governed by the land and forest law, and changing land categories is strictly prohibited. Since 1945, the DPRK government has closely adhered to these policies and enshrined them in law to protect existing arable land and to reclaim more land for agricultural production. In the late 1980s, arable land was well preserved as a result of strong economic growth and increasing crop productivity, and the loss of arable land that arose due to infrastructure development was offset by additional reclamation of tidelands for agriculture production.

In the early 1980s, the DPRK initiated a program to reclaim 300,000 ha of tidelands in western coastal areas for agricultural production use, and these activities are now being vigorously implemented.

Table 4.3 Agricultural land use change (unit: 10^3 ha)

Year	*1990*	*1993*	*1996*	*2002*	*2005*
Amount	2,038	2,087	2,103	2,103	2,042

Since the mid-1990s, the country was adversely affected by natural disasters such as floods and landslides, which caused the destruction of arable land. Infrastructure development, such as road construction, and urban and settlement area expansion, was also a large contributing factor to the loss of arable land. However, agricultural land use change has remained stable since 1990, as seen in Table 4.3.

Throughout the last few decades, there have been periods of economic difficulties and unstable food production on arable land. The crop production activities of individual farmers and small farm groups have expanded to hillside forest lands, which has not only contributed to the alleviation of food safety problems in rural areas, but has also led to deforestation and forest degradation.

There are no specific categories for grazing land. According to government policy on promoting livestock farming, forest land suitable for livestock grazing was changed to grasslands for grazing, but there is no statistical data to verify this information.

Water land covers a small proportion of the territory and changed slightly. Major changes or increases occurred in reservoirs for both power production and agriculture, natural lakes, and irrigation channels. Construction projects of naturally flowing waterways, such as the Kaechon-Lake Taesong and Paekma-Cholsan waterway, irrigate lowland arable land in the western and southern areas. Irrigation of these areas was once conducted using pump stations and power supplies.

In the DPRK, gross forest area decreased by an annual average of 0.088% between 1990 and 2005, but increased by an annual average of 11,200 ha from 2006 onwards. Forest land is divided into protected forest, economic forest, firewood forest, and grazing forest. Area change of state forest land by year can be seen in Tables 4.4 and 4.5.

Table 4.4 Forest land change (unit: 10^3 ha)

Year	*1990*	*1996*	*2000*	*2005*	*2011*
Area	9020	8906	8902	8900	8967

Table 4.5 Forest land change by management objectives (unit: 10^4 ha)

Year	*1990*	*1996*	*2000*	*2005*
Protected land	112.9	112.1	95	96.1
Timber forest	544	536.6	506.5	505.4
Economic forest	143.6	138	117.9	119.9
Firewood forest	19.6	31.4	33.3	40.3
Grassland	17	19.4	20.4	20.4

Major Drivers of Land Use Change

Economic Development and Economic Crisis

There are many underlying factors driving land use changes that have occurred in complex patterns. Land use in the DPRK has changed due to socialist economic construction in order to improve the livelihood of the people. Although the DPRK was classified as an agricultural producing country in 1945, it gradually developed into a socialist industrialized country, as many factories and enterprises were built. For example, many large-scale factories were built, such as the Kumsong Tractor Factory, Sungri Automobile Factory, and the Taean electrical factory, and a wide area of agricultural and forest land was converted into industrial land. Many coal mines were also developed and expanded, which gave rise to an increase in industrial land. From the mid-1950s to the 1970s, the country's economy thrived and was in a period of rapid development. This period was known as the Chonlima-era (which can be directly translated as *the horse that runs a thousand ri in a day*, and signifies rapid economic development). The country transformed from an agricultural state into a socialist economy, with over 50% of GDP output from the industrial sectors, far exceeding agricultural sector contributions. The policy on socialist cultural construction meant some land was converted for cultural and educational activities. The growing GDP and revenues were channeled increasingly to investment in the agricultural sector, which has increased farming through mechanization, all-irrigation system, and sufficient chemical fertilizer supplies. By the end of the 1980s, total arable land had increased as a result of urban expansion, and infrastructure and construction development.

During early the 1990s to mid-2000s, the socialist market collapsed and the DPRK entered a period of economic hardship, during which time land use change in the agriculture and forest sectors was badly affected by the economic crisis. There was further deforestation of land, and parts of degraded land were managed by implementing an agroforestry approach.

Population Growth

Population growth of the DPRK is dependent on food, energy, resources, and more land for farming and housing. The population growth rate has increased by 0.86% between 1990 and 2011. Research data suggests the total population was 20.96 million in 1990, 22.11 million in 1996, 22.75 million in 2002, and around 24 million by 2011 (see Fig. 4.1).

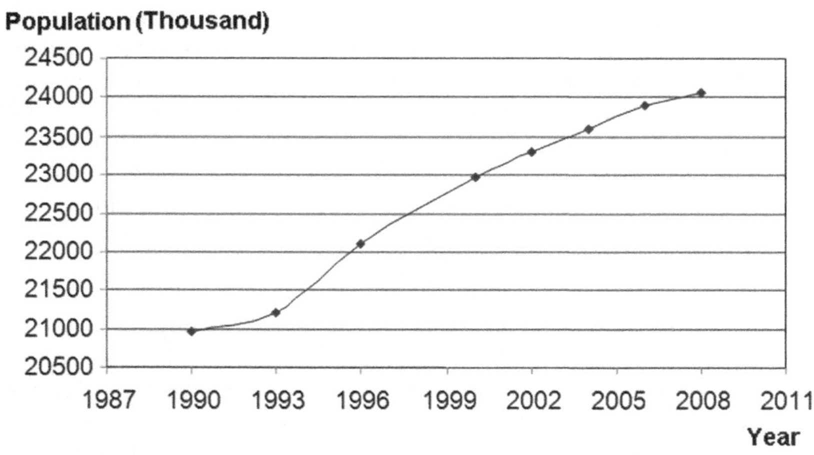

Fig. 4.1 Population growth in the DPRK (*Source* DPRK 2nd Communication on Climate Change [2012])

Urbanization

After the establishment of the DPRK, many new cities were built and expanded. For instance, Hamhung City, South Hamgyong Province, was a small industrial city that was made up largely of chemical factories, but was expanded to a large-scale industrial city with many chemical and machine factories. Similarly, Kaechon City was a small coal mine region in the 1970s, but developed to become a large-scale mining region in the 1980s. Urbanization was a measurement for economic development in the DPRK. The proportion of people living in urban areas was 60.2% in 1996 and 60.6% in 2008. The population of Hamhung City increased three times in the late 1950s due to rapid population growth, while the population of Kaechon City doubled in the 1970s. Coal is a basic fuel, and urbanization has caused increasing demand for coal in order to heat and cook food.

Infrastructure Development

Industry and agriculture development has seen an increase in transport roads, land area, reservoirs, and irrigation channels. Many roads and railways have been constructed and reconstructed, such as the Pyongyang-Wonsan highway, Pyongyang-Kaesong highway, Pyongyang-Hayngsan sight-seeing road, Manpo-Hyesan north railway, Ichon-Sepo railway, and the Wonsan-Onjong railway. Irrigation channels were constructed to cope with increased agricultural production such as the Yonpungho-Pyongnam, Two Thousand-Ri, and Sinkye-Miru irrigation channels.

Natural Disasters

Due to the adverse effects of climate change and increasing deforestation in recent decades, the DPRK faced severe natural disasters throughout the 1990s—some of the worst ever documented. The natural disasters were typically characterized by large frequency, scale, and intensity, which caused serious damage and required a large rehabilitation response.

Energy Shortage and Increased Energy Demand

Distribution and depletion of fossil fuels together with population growth have brought about energy shortages. Energy shortages are a contributing factor to deforestation, particularly in rural areas, due to over-harvesting of firewood. To overcome these shortages, the DPRK actively plants firewood forests, with a target of reaching 0.8 million ha of forest.

Economic difficulties have arisen due to economic sanctions and external pressures, which have resulted in reduced investment in the food production and land management sectors, which in turn has led to a continued drop in agricultural production and energy shortages. In mountain forests, total collected firewood increased from 3.2 million m^3 in 1990 to 7.2 million m^3 in 1996. As coal resources are further exhausted, forests in rural residences require more fuel, which is increasing rural dependence on forests.

Land and land use changes have been highly controlled by the state. Conversion from one land category to another was strictly regulated. In particular, arable land is highly regulated for farming purposes, and any changes to this category are unlikely to be approved. Since 1945, the DPRK has set and adhered to policies related to category changes. These policies are enshrined in law to protect the existing arable land and to help reclaim more land for agricultural production purposes. In the early 1980s, the DPRK introduced an initiative to reclaim 300,000 ha (good) of tidelands in western coastal areas and 20,000 ha of new land for agricultural production.

Reference

DPRK. (2012). *2nd Communication on climate change*.

CHAPTER 5

Forests and Forestry in the DPRK

I have worked with the FAO, IUFRO, and forestry workers in many countries for many decades. China's understanding of forests, forestry, regulations, policies, and governance closely aligns with that of the international academic community. From the perspective of management, we have a common understanding of sub-compartment, compartment, forest stands, and forest management schemes. However, in the DPRK, it seems that the basic views on forests and forestry are different to the international consensus. When engaging with DPRK counterparts on the topic of forests, I discovered discussions had to be specific and consider forest management implementation and restoration, so as to reach a consistent understanding. Irrespective of what had been agreed at the table in conference rooms, it seemed impossible to align our thoughts on this matter.

Definition of Forest

The DPRK has not yet developed terminology for its forest inventory. This explains why various references about forest cover in the DPRK provide competing and contradictory data. In various DPRK official sources, forest cover data was different and confusing too.

In the DPRK forestry community, forests have the following scientific definition: "forest refers to the ecological space where plants like shrubs,

L. Jinlong, *Crying Forests*,
https://doi.org/10.1007/978-981-16-9725-8_5

herbs and mosses, and trees grow and provide habitats for animals." Plants, animals, and microorganisms are organically integrated in the same environment and continuously undergo changes—this, as a whole, can be described as a forest. However, the mainstream definition among international forestry academics is "forest refers to an ecosystem, dominated by perennial plants, integration of environmental factors and biological factors." Thus, in the DPRK, environmental conditions, including temperature, sunlight, water, soil, and topography are considered as external factors that determine vegetation types and forest changes. Forest definitions in the DPRK can be traced back to the theory of phytogeocoenosis invented by Prof. Sukachiov V. N.,[1] a Russian expert in ecology and forestry.

Forested land is divided into tree-stocked land, non-tree-stocked land, and non-forested land dependent on degree of tree and plant cover. Tree-stocked land is defined as having a minimum DBH of 4 cm and minimum crown cover of 0.3. Non-tree-stocked forest land has very few tall trees and economic herbs due to tree cutting and other human activities, but areas with less than 0.3 tree crown cover can be converted into tree-stocked land. Non-forested land consists of facilities for forest management and production of forestry, firebelts, motorways, and railways in forests, field of seedlings, and storage fields of cut trees.

It is globally accepted that a forest should have a minimum area of 0.05 ~ 1 ha with 10 ~ 30% crown cover and minimum tree height of 2 ~ 5 m for the classification and assessment of forest resources. Working within these criteria, the DPRK has defined forest as a land with trees over 3 m tall, 30% crown cover, and a minimum area of 1 hectare for forest inventory. However, these criteria for forest inventory are rarely used by

[1] Sukachiov Vladimir Nikolaevich (1880 ~ 1967) used to be famous scientist from the former Soviet Union. He has mainly contributed to develop forest typology theory and establish phytogeocoenosis theory. He has made many academic achievements in several theoretical issues such as dendrology, meadow, phytogeography, paleobotany, geographic landscape, and biology. His forest type theory has a great influence in the East European countries and other socialist countries, including China and DPRK. After opening-up since the 1980s, China forest academia prefers to other theories developed by North America and European, and less and less foresters in China even know about the theory of Phytogeocoenosis, none of younger forest scientists study theory of Phytogeocoenosis.

the forest community, and I presume there is no professional nation-wide forest inventory to test these criteria. Generally speaking, whenever I asked *what is a forest?* The only definition received was Sukachiov's scientific definition of a forest.

Classification and Management of Forest Resources

In the DPRK, forests are classified into specially reserved forests, general reserved forests, and forests for production according to management objectives. Specially reserved forests and trees were identified at the sites and included revolutionary battle sites, historic sites permeated with anti-colonial sentiment, monuments of three great Generals of Mt. Paekdu and special objects, which were used to protect the surrounding forest environment from natural disasters. The main management seeks to maintain sound forest health by selecting and planting appropriate tree species and banning tree cutting, as well as increasing protection against damage from natural disasters, forest fires, and pests.

A general protected forest provides favorable living conditions and a good environment for conducting production activities. It can be broken down according to management objectives; a forest for protected areas, environmental protection, and academic research. Protected forest areas are forest districts chosen for conserving ecosystems and species or providing a habitable environment for people, with the main aim of conserving and expanding species, groups, communities, and natural environments. They include biosphere reserves, natural parks, natural reserves, plant reserves, animals reserves, and historic relic reserves. Forest management activities aim to improve the quality of forests by enhancing environmental services, such as natural rehabilitation of biodiversity conservation and forest restoration for soil and erosion control.

Protection forests can be classified into erosion control forests, large river forests, watershed protection forests (reservoir protection forests for hydroelectric power and irrigation, catchment basin protection forest), windbreak forests (windbreak for coast, paddy, and non-paddy protection forests), and scenic forests (city scenic forests, coast scenic forests, border coastal protection forests, traffic scenic forests, recreation forests).

Forests in academic research include scientific research forests and forests for university training. A production forest is divided into areas for timber, fuelwood, fiber or pulp, oil, wild fruit, honey, perfume, medical herbs, mushrooms, and seed collection.

According to the land classification status reported by the main management objective in 2005, protected forests accounted for 12.5% of forested land area, timber forest 66.8%, economic 15.5%, and fuelwood forest 5.2% (*Source* DPRK, 2012).

Forest Land

All forest land comes under state ownership in the DPRK. The area defined as forest land encompasses 89,020 km^2 (or 73.2% of total land area) of which existing forested area covers about 7.6 million ha. This is 0.4 ha of forest area per capita, which is lower than the global average of about 0.62 ha/capita.

Forest land is divided into monument forest land and other forest land. Forest land can also be classified into forest land with forests and forest land without forests. It is reported that there are 76.432 square kilometers of forest land with forests, however it actually is much less than that amount, but there is no official data available. Forest land without forests amounts to 8768 square kilometers.

Mountain forest land is divided into the following categories: special protection forest, general protection forest, timber forest, economic forest, and firewood forest for management. Timber forest makes up the largest total area. In 2010, the total area of protected forest was 9535 square kilometers, including sanitary wind forests, windbreak forests, traffic protection forests, water conservation forests, riversides forests, academic research forests, natural protection forests, scenic spots and cultural relics protection forests, border protection forests, and wildlife protection forests. Economic forests are used for fiber and binary raw material production, oil production, medicine, sour fruit, Party member fee, and spice production, covering an area of 11,816 square kilometers. Firewood forest is used for heating, cooking fuel, and agricultural reproduction, covering an area of 3.988 square kilometers.

Forest land is divided into industrial forests, economic forests, firewood forests, and protection forests (for further details see Table 5.1). Industrial forests are forests managed for the production of logs or for the processing of usable timber. Economic forests are used for producing oil, medicine, fiber, spice, magic, and fruit. Protection forests are there to protect biodiversity. Firewood forests are found in the countryside and used for growing grass.

Table 5.1 Forest area of different categories in 2005

Type of forests	*Annual area, km²*	*Area proportion, %*
Total	85,200	100
Industrial forest	50,902	59.7
Economic forest	11,816	13.9
Firewood forest	3,988	4.7
Protection forest	9,535	11.2
Land for forest	8,959	10.5

Source Approval of the United Nations development plan, prepared in October 2009

Forest Resources

Table 4.2 introduced forest resources by categories, dividing them into economic forests, non-tree-stocked land, and stocked land. Areas that contain economic plants have shrubs and economic herbs with a DBH of less than 5 cm. There are species of pepper tree prickly ash (Fagara schinifolia), European hazel, bush clover (Lespedeza bicolor), and Buxus in economic herbs, and herb species that belong to Insam, Mansam, bracken (Pteridium aquilinum), balloon flower (Platycodon grandiflorum), arrowroot, and mushroom.

Non-tree-stocked land means forest land with very few tall trees and economic herbs due to tree cutting, wood burning, pests, damage caused by domestic animals, and abnormal weather conditions. It can be converted from tree-stocked land to rare tree sites, bushes sites, grasslands, and harvested sites. Non-forested land also consists of areas for forest management stations, firebelts, motorways, and railways in forests, fields of seedlings, and storage fields of cut trees. Thus, non-forested land is necessary in forest management and forestry activities, but it doesn't produce forest products and doesn't have a role in environmental protection. Non-forested land also has land that is unfavorable for growing forest plants due to geographical and geological conditions including land that lies above the forest distribution line and abnormal weather. Forested land is divided according to forest cover as can be seen below (see Table 5.2).

Forested land is divided into three types of forest in accordance with component of species; coniferous forest, broad-leaf tree forest, and coniferous-broad-leaf mixed forest. A coniferous forest accounts for

Table 5.2 Forested land resources (Unit: 103 ha)

Total	*Stocked*	*Non-stocked*	*Non-forest*
87,170	69,120	14,530	3,520

Source Central Statistic Bureau (October 2011)

Table 5.3 Area ratio by tree species of tree-stocked land

Name	*Oak*	*Larch*	*Pine*	*Deodar (fir, spruce,* Pica Koraiensis*)*	*Pine nut*
Ratio (%)	29.5	17.5	12.7	8.2	5.8

Source DPRK 4th Report on Biodiversity (2011)

41.9% of tree-stocked land, a broad-leaf tree forest 35.6%, and mixed forest 22.5%.

In 2005, the area of natural forest covered 5,790,000 ha, which accounted for more than half of total area, and planted forest area totaled 1,824,000 ha. The main forest tree species were pine, spruce, fir, oak, larch, Korean pine, birch, and aspen trees. Table 5.3 provides an area ratio by different tree species of tree-stocked land. Average standing stock volume of forested land was 68 m^3/ha, which was much lower than the global average of 100 m^3/ha at the time.

Non-timber forest resources such as wild fruits, mushrooms, and medical plants are significant additional food and income sources for local industries and rural livelihoods. In the last few decades, these non-timber forest resources have been collected by local people beyond their regeneration capacities resulting in remarkable resource decline.

Forest Stocks

Total forest growing stock volume was 330,453,000 m^3 in 2010 and average stock volume per hectare amounted to 47.8 m^3. The present stock per hectare is very low—less than half of the world average of 110 m^3/ha. The low forest growing stock volume is attributable to high forest resource exploitation and low efficiency in afforestation and forest management.

Timber resources are widely used in various sectors of the economy, both directly and indirectly, including the construction, mining, building materials, and railway industries. For example, almost all prop in the coal sector and 60% of tie in the railway sector rely on timber resources, which shows how increases in production in these sectors have a direct link with timber production.

Non-timber Forest Resources

Non-timber forest resources include tree oil, medicinal, edible herbs, wild fruit, and bee plant resources. These resources are valuable natural raw materials in the pharmaceutical and foodstuff sectors and provide a source of income to local people in mountainous areas. Table 5.4 depicts some non-timber forest resources in 2010. Table 5.5 lists the main species found in non-timber forest resources in the DPRK.

April, May, September, and October are designated herb growing months, and it is encouraged to grow or gather medicinal plants. There are roughly 900 species of medicinal herbs in the country and about 1,700 species are used in the Koryo medicine field (Korean traditional medicine). As of 2011, the cultivation area of medicinal plants was approximately 43,000 ha and protected areas covered about 93,000 ha (*Source* DPRK 4th Report on Biodiversity, 2011).

So-called forest conservation groups that consisted of retired workers and housewives were formed in order to provide the local economy with non-timber forest products. These groups were allowed to cultivate designated hill areas, called sotoji (small pieces of farmland), when there was inadequate food provision from the central government. The emergence of sotoji was a contributing factor to forest destruction during the 1990s and also contributed to the food crisis of the 2000s. The central government made several attempts to prevent the use of sotoji

Table 5.4 Main non-timber forest resources

Medicine		*Edible herbs*		*Mushroom*	
Area (ha)	*Production(t)*	*Area (ha)*	*Production (t)*	*Area (ha)*	*Production (t)*
339,861	2,765	461,700	5251	201,500	1,700

Source General Forest Resources Inventory (2010)

Table 5.5 Species of non-timber forest resources

Classification	*Species*
Fiber and pulp tree	Over 100 species including: poplar tree (*Populus*), *Metasequoia glyptostroboides*, deodar, trifoliate pine, lime tree (*Tilia amurensis*), bush clover (*Lespedeza bicolor*) and bamboo tree
Wild fruit tree	40 species including: chestnut, apricot, tara vine, Mongolian oak, *Rubus crataegifolius*, *Crataegus pinnatifida*, *Schizandra chinensis*, and jujube tree
Oil tree	60 species including: walnut tree, pine-nut tree, *Juglans cordiformis*, ginkgo tree, *Evodia daniellii*, *Fagara schini floia*, hazel tree
Medicinal tree	Over 30 species including: *Schizandra chinensis*, juniper tree (*Sabina chinensis*), hawthorn (*Crataegus pinnatifida*), Amur cork tree (*phellodendron amurense*), matrimony vine (*Lycium chinense*), ginkgo tree, strawberry tree, *Schizandra chinensis*, *Sabina chinensis*, *Crataegus pinnatifida*, *Phellodendron amurense*, *Lyctum cinnese*, Ginkgo, *Rubus crataegi folius*, *Echinopanax elatum*, and *Aronia melanocarpa*
Honey tree	Approximately 170 species including: a lime tree (*Tilia amurensis*), an acacia (*Robinia pseudo-acacia*), a bush clover (*Lespedeza bicolor*), and a chestnut tree (*Castanea*)
Edible wild plants	Approximately 300 species including: *Aralia elata*, *Acanthopanax sessiliflorum*, and *Kalopanax pictum*
Aromatic plants	Approximately 60 species including: an aromatic rose, lilac *(Syringa dilatata) Magnolia obovata*, magnolia (*Magnolia sieboldii*), deodar, and *Thuja koraiensis*
Wood rosin	Approximately 10 species including: a lacquer tree, a pine tree, and deodar

but ultimately failed, as local people became dependent on these lands for wood, cooking, and heating. Thus, the local governments that had overall responsibility for these forests were unable to impose strict regulations to restrict the use of sotoji, and so usage continues even today. International aid organizations, such as ICRAF and CIDA, renamed sotoji as "user group," a name that fascinated global donor organizations due to its association with decentralization, democracy, empowering communities, and detachment from the authoritarian government. With support from international donors, agroforestry development in sloping lands has become "state policy," but I could not tell you what the exact meaning of this policy is in practice. In 2009, the DPRK enforced a policy that prevented the prohibited felling of trees and cultivation of forest land, forcing people to sign a document called the "Contract for Forest Protection." However, the effectiveness of this policy is highly questionable, as local people need food and fuelwood to survive.

Forestry Legislation and Administration

Forestry Legislative Framework

The DPRK has adopted laws and regulations with regard to the protection and management of forests since the country's liberation, and continues to amend and adjust provisions as required. The laws related to the protection and management of forest are described as follows:

1. "Law on Land of the DPRK" was legislated on April 29, 1997 and amended on June 16, 1999.
2. "Law on forest of the DPRK" was legislated on Dec. 18, 1992, and amended and adjusted on Feb. 4, 1999, Sep. 10, 2001, Oct. 18, 2001 and Sep. 2005.
3. "Law on environment protection of the DPRK" was adopted on April 9, 1986, and amended and adjusted on March 4, 1999, and July 24, 2000.
4. "Law on useful animals' protection of the DPRK" was legislated in 1998 and amended and adjusted on July 24, 2000.
5. "Law on natural reserve of the DPRK" was legislated in 2010.
6. "Law on land plan of the DPRK" was legislated on March 27, 2002.
7. "Law on control of land and environment protection of the DPRK" was legislated on May 27, 1998.
8. "Law on protection of scenic spots and living monuments of the DPRK" was legislated on Dec. 13, 1995, and amended and adjusted on January 14, 1999.

The Law on Forest of the DPRK was enacted on December 11, 1992, the first separate piece of legislation for forest management and protection that abandoned its traditional forest management style in favor of territorial integrity and timber production. Enactment of this law showed how official perceptions of the environment and forests began to change, taking concrete steps toward forest protection (Kim & Ali, 2016). This law emphasized forest management under the unified guidance and control of the state, in comparison with the focus of the earlier Land Law. This legislation suggested that the central government sought to reassume the forest management role that it had handed over to the local government (Kim & Ali, 2016). The Forest Law was amended and supplemented four times, in February 1999, September and October 2001, and August 2005. These frequent revisions show that the DPRK put forestry in its political agenda to address severe environmental problems. In particular, the revised Forest Law of 2001 sought to address several grave problems, such as illegal timber exports and the expansion of slash-and-burn fields. With the expansion of South Korea and the international community's support for forest restoration, the Forest Law was reinforced in 2005 to include legal provisions governing foreign tree species.

Forestry Administration

In the DPRK, at the central government level, the Ministry of Land and Environment Protection (MoLEP) and Ministry of Timber Production (MTP)/Ministry of Forestry (MF) are in charge of developing and managing forest land in favor of timber production and processing. They manage, supervise, and control the forest management under the unified guidance of the Cabinet. The department of forest administration, MoLEP, is responsible for giving unified guidance, managing national forests as a whole and controlling the remainder of the country's woodlands in order to supervise their territorial integrity. MF is in charge of the forests that produce timber for industry needs, and responsible for the establishment, cultivation and utilization of the cyclic felling zone under the supervision and control of the Department of Forest Administration of MoLEP. MoLEP not only manages all businesses related to the use of national land and natural preservation, but also the fishery and agricultural assets not covered by their respective ministries. The Ministry of City Administration is in charge of forests in urban areas. The Ministry of Agriculture is responsible for certain parts of forest land for farming materials, fuelwood, and livestock.

Both ministries have a well-ordered administrative management system down to the provinces, cities (counties), and ris. Table 5.6 shows how the DPRK allocates forests to different administrative organizations in accordance with its management style and goals vertically. The Bureau of Land and Environment Management is an affiliate of the Ministry of Land and Environment Protection at the provincial level, which supervises land preservation and forest reservation and is in charge of forest administrative

Table 5.6 The Forest Administrative System in the DPRK

Administrative structure	*Agency in charge of timber production*	*Agency in charge of land and environment management*
Central government	Ministry of Forestry; Bureau of Forest Production Management	Ministry of Land and Environment Protection; Bureau of Forest Management
Province	Bureau of Forestry Management; Forest Products Station	Bureau of Land and Environment Management Protection; The Forest Development Design Station
County	Industrial forest division; Forest production division; Propaganda division	Department of Land and Environment Protection Management; Forest management; station Silviculture station
Village	Silviculture operation	Forest supervisor office

Source Park Kyung-Seok et al. (2008)

services. Through the Forest Development Design Station, the bureau sets up policies for the management of forests and nurseries suited to the workforce of each province and county, and is responsible for the labor and the budget needed to execute its policies. The affiliates of the Ministry of Land and Environment Protection at the county level are the Department of Land and Environment Protection Management, the Forest Management Station, and the Silviculture Station.

Each province has a division of forest management under the department of land and environment protection and management at a provincial level, and each county has a division of forest management boards and industrial forest at the county level. The Forest Management Station was set up under the Ministry of Forestry and then transferred to the Ministry of Land and Environment Protection. It is responsible for the development and protection of forests and plays a crucial role in educating forest supervisors with regard to preventing forest fires. The forest Silviculture Stations in every county manage forests and organize afforestation. The Silviculture Operation Office at the village level is responsible for tree nurseries and the planting of trees. Forest supervisors manage the timber harvesting and production of forest products and even watch

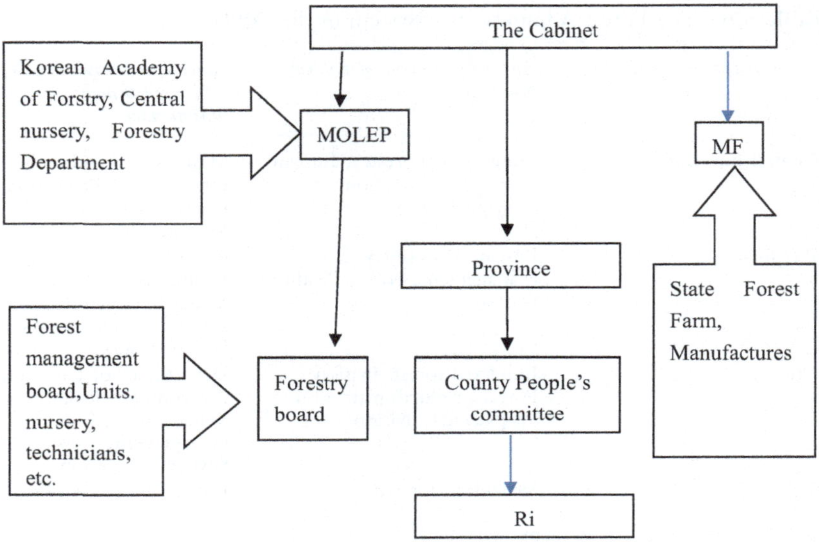

Fig. 5.1 Forest Vertical Administrative Chart in DPRK

out for forest fires from sentry posts, and have a role in educating the public about forestry policies, afforestation, and forest conservation work. There are several branch offices of supervision and afforestation, nursery stations, and working stations for erosion control in county-level forest management boards (see Fig. 5.1).

For forest land near individual factories, schools, and social organizations in rural areas, the protection, management, and product utilization of certain areas are allocated among those organizations for their specified purposes with permission by authorized agencies.

Forest Management Unit

All the agencies and units belonging to government including cooperative farms, enterprises, universities, researching institutions, schools, hospitals, and land user groups can manage and use some forest land as production land during certain periods if approved by the country. These agencies and units are given supervision and control of MoLEP in using lands. For example, a forest for the timber industry is used permanently with the cyclic felling method and forests for farm and factory are used with reapproval once every 5 years on the basis of understanding and assessment of forest utilization. The tenure of the experimental forest is generally 10–50 cycle years in accordance with the research objective.

Communities in local areas can possess about 90 m^2 of land per household and produce agricultural products necessary for their lives from the land. Approximately 2.5 million m^3 of timber and 7.2 million m^3 of fuelwood have been harvested every year since around 2010. There has been no change in area of forest land over the last century, but during the last few decades the forest coverage has greatly declined due to recurrent natural disasters and economic difficulties. This has resulted in low forest growing stock, loss of biodiversity, and the deterioration of forest ecological functions, which was followed by soil erosion, landslides, continuous flooding, and weak environmental stability for agricultural production.

Natural Reserves

The DPRK has rich biodiversity and high landscape value. The DPRK set high protective valuable areas and habitats of bio-species as natural reserves, which have been established and reestablished in line with the requirements of the developing situation. Historically, nature reserve development can be seen as:

- The DPRK promulgated "decree on maintaining treasures, historic spots, scenic beauties and living monuments" in April 1946 and listed major natural monuments and historic relics in 39 regions as protected areas.
- In February 1959, the Cabinet established the first ever protected areas.
- In 1976, protected areas in the country were rearranged.
- By 1994, 6 natural reserves (one biosphere reserve), 14 plant reserves, 14 animal reserves, and 6 sea-bird breeding reserves had been established.
- By 2003, a total of 127 natural reserves were established including: one biosphere reserve, 21 natural parks, 8 natural reserves, 24 landscape reserves, 28 animal reserves, 6 sea-bird reserves, 12 seasonal-bird tideland reserves, 9 marine resources reserves, 6 sea reserves, and 12 freshwater reserves.

- By 2012, 233 natural reserves had been reestablished or newly established, including: 3 biosphere reserves, 6 natural reserves, 27 plant reserves, 31 animal reserves, 38 marine resources reserves, 48 living monument reserves, and 95 natural parks.

The protected areas accounted for 5.7% of the total territory prior to 1990, increased to 6.1% by 2000, and has risen to over 7% in 2009. Protected forests include natural reserves within forests and ecological environment protection forests (a bank-protecting forest, a reservoir protection forest, an erosion control forest, and a hygienic and ornamental forest).

Since 2010, the protective value of natural reserves has diminished due to constant pressures on them. These constraints can be divided into direct impact, indirect impact, natural, and anthropogenic. Typically, they include illegal resource exploitation such as forest fires, hunting and timber harvesting, loss of habitat by natural resource over-use, forest pest outbreaks, invasive species, loss of soil and water resources around the protected areas, climate change, and pollution of water quality.

Importance of Forests to the Economy and Environment

Contribution to the Economy

Forests play an important role in economic development in the DPRK, with many sectors reliant on forest resources for raw materials needed for production. From 2000 to 2009, annual wood logging in forest areas amounted to 2,730,000 m^3, of which, 700,000 m^3 was used for panel wood production and 300,000 m^3 was used for railway sleeper production (*Source* MoLEP, 2010). In addition, hundreds of thousands of cubic meters of wood were used annually for construction and building materials. Forest timber production largely occurs in the northern regions, as does construction and mining.

Non-timber forest resources are also important for the development of local industry and improvement of livelihoods. Local foodstuff factories and pharmaceutical plants rely on forest resources for 50% of raw materials, this has come about as a result of government policy that seeks to develop local industry by localizing raw material production.

Non-timber forest resources have become a valuable income source for local people. Although locals increasingly depend on natural foodstuffs, people living in mountainous and hilly areas tend to rely on different wild fruits and mountain herbs as their main source of income. For example, locals living in mountainous and hilly areas often harvest many tons of non-timber forest resources, especially in the following areas: Unsan, Kujang, and Hyangsan County, North Pyongan Province, and Gaechon and Tokchon City, and South Pyongan Province. Now, a number of different non-timber forest resources are gradually increasing in local markets.

Rural Energy

In the DPRK, forests have been a valuable source of energy for local households for centuries. In rural areas, energy is used for cooking and heating, and the average amount of firewood needed per family is 5 ~ 7 m^3 per year, which totals 7,000,000 ~ 8,000,000 m^3.

In the 1980s, the state provided locals with energy including electricity and coal. Forests account for about 60% of total rural energy, but since the 2000s, this has risen to 80% due to serious energy shortages (for details see Table 5.7). Forests have become an important source of fuelwood in rural areas due to a decline in alternative energy supplies. Although cooperative farms, factories, and other organizations were allocated firewood plantation forests, these designated forests did not reach required utilization standards due to poor management, so firewood collection took place beyond these appointed areas. Average fuelwood demand in rural areas could have reached 12 to 14 million cubic meters every year had alternative energy sources not been utilized. In fact, as firewood collection largely occurs in an uncontrolled and illegal manner, there is very little statistical data, but one thing is clear that over-collection of firewood in many hilly forests has led to severe forest degradation.

Table 5.7 Amount of firewood required at 2009 (unit: m^3)

Area	*Heating*	*Cooking*
Urban	919,000	984,000
Rural	1738,000	1774,000

Water and Soil Conservation

Forests play an important role in water and soil conservation as the DPRK has a lot of rivers and streams, and comparatively high precipitation. There are about 10,000 watersheds in DPRK, all of which are surrounded by woods and many are surrounded by arable lands. 40% of areas under cultivation are situated on sloping land. In recent decades, natural disasters have brought about damages to rural areas due to a malfunction of forest ecosystems for environmental protection such as soil erosion control and regulation of water run-offs.

In recent times, deforestation has increased in severity and natural disasters resulting from climate change and deforestation are occurring more frequently. Therefore, water and soil conservation in woods are an issue of existence and environment.

Biodiversity and Other Environmental Values

Biodiversity is the variability of life on earth measured at the ecosystem level. In particular, forests are key to maintaining biodiversity and play a vital role in social, cultural, and economic development and ecosystem conservation. Forest ecosystems are critical in conserving biodiversity in the DPRK, where most land is densely covered by forests.

Recreation and Other Spiritual and Cultural Values

The DPRK's forests are also increasingly valued for their beauty and recreational potential. Major mountains such as Mt. Kumgang, Mt. Myohyang, Mt. Kuwol, and Mt. Chilbo are considered protected areas due to the rich biodiversity and beautiful landscapes, and also attract many domestic and international tourists yearly. Due to high annual footfall, the DPRK intends to develop similar zones to attract even more visitors.

Although the benefits of forests are clear from an economic, social, and environmental perspective, they are often poorly managed. Poor forest management is closely related to inefficient forest resource, service, and function management. Therefore, it is necessary to conduct further studies on forest services and functions, to determine how best to improve management in a sustainable manner that enhances socioeconomic and environmental benefits.

References

Central Statistic Bureau. (2011). *Statistical report on achievement of MDA in DPRK*.

DPRK. (2011). *4th report on biodiversity* (English).

DPRK (Democratic People's Republic of Korea). (2012). *Environment and climate change outlook*.

Kim, R. E., & Ali, S. H. (2016). Green diplomacy—An opportunity for peace-building? *Environmental Policy and Law, 46*(1), 86–96.

MoLEP. (2010). *Gereral forest resources inventory*.

Park Kyung-Seok, et al. (2008, December 8). *A study on the trait of the forest system of the central and local governments and its management system in The DPRK*. Korea Forest Research Institute, Research Report.

CHAPTER 6

Change of Forest Cover and Quality

In the DPRK, forest land accounts for 80% of total land area and water body just 6%. Management of mountains (shan) and water (shui) has long been regarded as a high priority area for addressing environmental concerns (including climate change and biodiversity loss), promoting social-economic development and increasing food security. In the DPRK, landscape can be interpreted as the integration of "shan" and "shui."

Changes in Forest Cover and Growing Stock

The DPRK's population has doubled since the 1940s. Industrial and infrastructural developments (such as roads, irrigation canals, and reservoirs) have resulted in nearly 500,000 ha of forest land being officially reclaimed for use as residential, industrial, and riparian areas in addition to farmland (Table 6.1). Non-forested land area has decreased and increased overtime, comprising a small share of total forest area (Table 6.2). Difficult to analyze, these areas consist of: (1) infrastructure and facilities used for forest management and production (e.g., fire prevention belts, motorways, railways, nursery land, and timber storage land) and (2) land that is unfavorable for growing forests (e.g., steep rocky mountains and land with abnormal weather conditions).

Forested land areas began to decrease due to severe repercussions of war beginning in 1948. These areas increased to a peak of 8,435,000

L. Jinlong, *Crying Forests*,
https://doi.org/10.1007/978-981-16-9725-8_6

Table 6.1 Changes in forest land by category (Unit: 10^3 hm^2)

Year	*1948*	*1954*	*1962*	*1979*	*1990*	*2000*	*2010*
Forests lands	7849	7557	8328	8435	8133	7554	6912
Denuded forest lands	1007	–	473	338	432	870	1453
Non-forested lands	358	–	214	197	380	405	352
Total	9214	–	9015	8970	8945	8829	8717

Table 6.2 Changes in forested land area according to management purpose (Unit: 10^3 hm^2)

Management purpose	*1972*	*1979*	*1990*	*2000*	*2010*
Timber	5552	5375	5440	5090	4873
Economic trees	1126	1199	1436	1182	1235
Fuelwood	81	151	196	399	427
Protection, recreation, and unclarified	–	1710	1061	883	377
Total	–	8435	8133	7554	6912

hm^2 in 1979 before slightly decreasing after the collapse of Eastern European socialism. Beginning in 1990, the denuded forest land area rapidly increased, reaching 1,453,000 hm^2 in 2010 according to my estimates, deduced with the help of foresters in the DPRK (Tables 4.2 and 4.3). Data from satellite images provided by the Korea Forest Research Institute shows that the area of degraded forests in the DPRK is 2.84 million ha out of a total forested area of 8.99 million ha (Park et al., 2013). The forest cover figures published for the DPRK by the Forest Resource Assessment and FAO (FAO/FRA, 2010, 2020) forecasted a rapid decline in forest cover from 8.20 million hectares in 1990 to 5.67 million hectares in 2010, representing a deforestation rate of more than 2% per year, that has continued declining in recent years. An official government report released when the DPRK asked for international assistance estimated the nation's forested area as between 7,750,000 ha and 9,850,000 ha (Park, 2010).

Table 6.2 provides another indicator to assess forest cover changes from 1972 to 2010 in accordance with management purposes. Economic tree plantations remain quite stable. The forest area used for timber products managed directly by the Ministry of Timber Production reduced by

Table 6.3 Forest growing stock in the DPRK

Year	*1948*	*1962*	*1979*	*1990*	*1996*	*2000*	*2010*
Stocking volume (106 m^3)	600	466	515	526	499	427	330

Table 6.4 Top three countries for deforestation

State	*Total area (10^6 ha)*	*Area of deforestation 1990–2000 (10^3 ha)*	*Growing stock 2010 (10^6 m^3)*
Brazil	832	5332	126,221
Indonesia	181	2412	11,343
DPRK	12	1022	360

approximately 679,000 ha between 1972 and 2010. However, the forest area used for fuelwood expanded to reach four times its original size.

Growing stock is another indicator for assessing changes in forest cover. Table 6.3 shows a variable curve after 1948. In the 20 years since 1990, the DPRK has lost 37% of its growing stock, while per ha growing stock indicates a decrease of 78 m^3 to 58 m^3 after deducting the area used for economic tree plantations. This is a decrease of approximately 26%, showing a great decline in forest quality.

Overall, fewer DPRK forests were lost to official agricultural practices than those lost to illegal practices. Forest acreage was reduced from 8.1 million ha to 6.9 million ha from 1990 to 2010. Meanwhile, growing stock decreased from 526 million m^3 to 330 million m^3 (nearly 3% annually). The DPRK ranks as the third worst country for deforestation after Brazil and Indonesia. When considering the ratio of total forest area to total territory in each country, DPRK deforestation actually seems much worse (Table 6.4). Among the top 10 countries for deforestation, the DPRK is now experiencing the most dramatic decreases in stocking volume.

Forest Biomass and Forest Carbon Stock

Forests, a main part of the global carbon cycle, are not only affected by climate change, but also affects climate change. The assessment of forest biomass and carbon stock is significant to calculating certified emission

Table 6.5 Forest biomass and carbon stock by forest growth rate

Year	*1990*	*1996*	*2000*	*2005*
Total forest biomass (t)	451,360,000	449,360,000	384,170,000	386,880,000
Total carbon storage (t)	225,680,000	224,680,000	192,080,000	193,440,000

reductions (CERs) by the carbon sink. Forest biomass and carbon storage are currently assessed based on IPCC guidelines, and so it is important to note assessment results.

Biomass and carbon storage can be calculated by applying average factor prices (0.7 tons of biomass on ground, 0.9 tons of total biomass, and 0.5 tons of carbon are the biomass equivalent to 1 m^3 of forest growth accumulation). Table 6.5 shows the carbon storage change from 1990 to 2005 by forest accumulation.

Management Purpose

Area of non-wood land gradually increased from 1979 to 2000, but decreased in 2010. Land use changed between 1972 and 2005 in accordance with management changes. The land use change in timber forests, forests of economic value, and fuelwood forests are as follows (see Table 6.6).

The table above shows that in the period 1972–2005, the area of fuelwood forest expanded, while timber forest declined. The reason for this change in forest land use is due to the DPRK's forest policy that values forest protection and supports a transition to sustainable forest management.

Table 6.6 Land use change by management purpose (km^2)

Year	*1972*	*1979*	*1990*	*2000*	*2010*
Timber forest	55,520	53,750	54,400	50,900	48,730
Economic trees forest	11,260	11,990	14,360	11,820	12,350
Fuelwood forest	810	1,510	1,960	3,990	4,270

References

FAO. (2010). *Global forest resource assessment report.*

FAO. (2020). *Global forest resource assessment report.*

Park, J. H. (2010). A research on the actual condition of North Korea's forest devastation through MODES satellite image. In *The north and south's cooperation for North Korea's forestation* (pp. 73–84). Institute for Peace and Unification Studies at Seoul National University.

Park, K. S., Lee, S. Y., & Park, S. Y. (2013). Changes in North Korea's forest policy. *North Korean Review, 9*(1), 20–29.

CHAPTER 7

History of Deforestation and Forest Degradation in the DPRK

In 1945, forest resources covered up to 80% of DPRK land surface (Park et al., 2013). The amount of forest coverage in the DPRK has fluctuated several times over the course of a century. For instance, it decreased during the period of Japanese colonization, increased slightly after liberation, decreased again during the Korean War, subsequently increased because of socialist economic progress, and then dramatically decreased in 1990, when Eastern European socialism collapsed. Between 1990 and 2010, the amount of forest acreage in the DPRK decreased by 2.2 million ha, while the volume of growing stock continued to decrease by approximately 3% per year. This was partially due to the copious amounts of fuelwood used for cooking and heating as well as the practice of widespread slash-and-burn cultivation for food crops. These reverses in forest quantity and quality required comprehensive measures, including favorable international relations, political and institutional reforms, the decentralization of forest management, and proper technical support for local communities. The forestry issue is complex and, therefore, is no longer solely the prerogative of the forestry sector, but is embodied in political, social, cultural, economic, environmental, and other broad development issues. Without settling down all these complicated issues, forests in the DPRK have undoubtedly degraded and deforested continuously.

L. Jinlong, *Crying Forests*,
https://doi.org/10.1007/978-981-16-9725-8_7

Historic Review of Forest Coverage Over One Century

The Korean Peninsula has been inhabited by humans for approximately 5,000 years; some documentation suggests this may be even longer. Culture on the peninsula was an integration of the farming and settlement culture of the Han Chinese as well as the grazing and nomadic habits of Mongolian civilization. Sparsely populated mountain regions tend to be characterized by common-pool resources. In the most densely populated region, most forests of the Korean Peninsula were owned by landlords and rich farmers, while high mountain areas contained a small proportion of temple-owned forests, extended family forests, and some community forests.

Japanese Colonization

Japan was the first Asian nation to adopt Western concepts of industrialization, science, technology, modern education, lawful society, and the practice of allocating raw materials to support such a society. Japan superseded China as the major regional power in 1890. The Korean Peninsula was under Japanese control from 1905 to 1945.

Japanese colonization entailed the establishment of formal institutional arrangements and forest-based legislation from what was considered a scientific perspective. However, the policies were also designed to use the forest as a significant resource for expanding Japanese imperialism. The Forest Administration Office was established in 1906 and expanded in 1911. Several legal devices were developed during this time, including the "Code on Forest" (June 1911), "Code on Registration of Korean Real Property" (1912), and the "Edict on Korean Forest Land Investigation" (1918). Japanese forestry education at the university level was established using curriculums from Europe and the United States, where forestry science originated. Many Korean scholars were sent abroad, including to Japan and Germany, to study forestry science, before returning to Korea to establish forest research and education programs. Japanese colonialism thus facilitated the entry of the Korean Peninsula into the era of modern forestry.

However, many virgin forests rapidly disappeared due to an increased demand for construction (particularly for railways) and resources for the Japanese invasion of China, Southeast Asia, and other Pacific Regions.

DPRK estimates from 1907 to 1925 indicate that the watershed regions at the Amnok and Tuman rivers (where easy access was gained) experienced a 50% loss of virgin forest areas through such harvesting practices; approximately 35 million m^3 of timber was harvested (Collection of Korean Geography, Vol. 10). In regions with limited accessibility (e.g., areas around the Taedong and Rimjin rivers), only high-quality timbers were harvested, including those from species, such as Picea, Abies, and Pinus koreansis.

While likely overestimated, sources indicate that average annual timber extraction amounted to more than 748,000 m^3 from 1933 to 1938 (Collection of Korean Geography, Vol. 10). Large amounts of fuelwood were required to supply heating for troop encampments and construction sites. Non-timber forest products (NTFPs; e.g., pine resin, medicinal herbs, and wild plants) were also harvested at high rates, including 476 tons of mushrooms, 92,765 tons of charcoal, 4,936 tons of bark from oriental oak trees, 27,601 tons of wild fruit, 11,058 tons of wild plants, 2,346 tons of medical herbs, and over 10,000 tons of pine resin each year from 1933 to 1938.

Overall estimates indicated that 11% of all virgin forests were converted to non-wooded lands, while about 14% of all forested lands on the Korean Peninsula experienced severe degradation.

Land Reform and the Korean War

The end of the Second World War marked the end of Japanese rule over the Korean Peninsula. However, with capitalist American forces on one side and socialist Russia on the other, the Korean Peninsula became an ideological battlefront. It was thus divided into two countries (i.e., the DPRK in the north and Republic of Korea (ROK) in the south).

As Korea was a socialist country, reforms were implemented so that all land was nationalized. The "Decision on land problem" was enacted on October 16, 1945, while "the Agrarian Reform Law of DPRK" was enacted on March 5, 1946. All forests have been nationalized since that time. No records are available to indicate what happened as a result of these land reforms, but forests do not typically receive much protection in transitional periods (Liu, 2009). Chinese sources indicate that landlords and wealthy households with relatively large areas of land took measures against these reforms, including creating forest fires. Poorer residents who were worried about forest security may have harvested areas in exchange for cash (Liu, 2009).

State-level forest management and supervisory institutions were established based on "The decision on forest land management by the Provisional People's Committee of the DPRK" which was enacted on June 4, 1946. The Forest Science Research Institute (now called the Academy of Forest Sciences under MoLEP) was established on December 29, 1948. Managerial and administrative institutions that supported the nationalized forests were suspended during the Korean War. The DPRK established rules for forest management practices and economic organizations on January 27, 1953. Thus, subordinate agencies attached to state-level forest management and control systems were also established. As such, a uniform forest management and administrative system spanning both state and local-levels was created. These still function in the DPRK with only slight transformations.

In December 1948, the DPRK conducted its first comprehensive forest inventory after establishing the Forest Inventory Working Group (now under MoLEP). It concluded that 9,213,600 ha were forest lands, 7,849,000 contained forests, and 138 million m^3 were comprised of growing stock. However, based on the national inventory data, forest standing volume in Northeast China, adjunct to DPRK forest areas is between 83–95 cumber meter per ha during 1990–2010, as these forests have been heavily harvested before (Liu et al., 2015). I believe the amount of standing volume was significantly underestimated, which might be no less than 600 million m^3 of standing volume at 1948 in the DPRK. Deforestation rapidly accelerated during the Korean War (1951–1953), at which time 291,622 ha of forest and 2,471,000 m^3 of timber stock were lost.

Period in the Socialist Countries Group

The DPRK closely allied itself with other socialist nations after the Korean War. The Ministry of Timber Products was then established to accelerate both industrialization and modernization. The ministry oversaw the management of forest-rich regions, which amounted to approximately 2 million ha of the remaining virgin forests used to maximize timber production beginning in the late 1950s. Timber production teams were organized during the early 1960s to conduct logging and afforestation operations in most of the remaining DPRK forest areas. Requirements stipulated that all logging would be performed sustainably. In particular, this involved the cyclic felling method. However, as proven in China, implementation was unlikely.

The DPRK adopted a heavy-industry prior development strategy to overcome domestic struggles in the 1960s. It also established a self-reliance strategy in response to the external economic crisis, thus consolidating the central government command economy to ensure security. Relatively robust economic development during the 1970s provided legitimacy to these strategies. This resulted in the Zhuche ideology being fundamentally embodied into the DPRK's political, economic, and governmental systems. This is still extant under so-called unshakable determination. DPRK policies greatly contributed to afforestation between 1960 and 1990 as well as the deforestation that occurred after 1991.

Afforestation occurred through two major means (i.e., public campaigning and an afforestation program). In March 1947, the DPRK officially announced April 4–10 as its Week for Planting. Millions of laborers were mobilized for this purpose, while politicians showed their desires to turn green mountains and hills into thickly wooded "treasure mountains" and "golden mountains." I was not able to find any evidence that the planting campaign convincingly contributed to afforestation. However, it may have simply been incorporated into the afforestation program itself.

The DPRK's afforestation program was successful in changing the land cover. In 1952, the success of Stalin's afforestation program was mimicked to create windbreak shelterbelt forests through massive efforts in the Northern highlands. This was intended to reclaim thousands of hectares of abandoned land for use in growing potatoes.

On March 4, 1960, the decision was made to organize and mobilize all people to engage in economically valuable afforestation. Through a so-called all-inclusive mass movement, around 1.5 million ha were reported to have been afforested with economically valuable trees by 1990. However, approximately 443,000 ha of cash tree plantations were increased from 1962 to 1990 (Table 7.1). Areas containing oil-tree and wild fruit-tree plantations also increased from 1962 levels, but areas containing fiber and pulp plantations were reduced. The area of economic tree plantations accounted for 11.6% of the total forest in 1990.

Another remarkable development during this phase was the creation of plantations for fuelwood, which were afforested in highly populated

Table 7.1 Areas Containing Economic Tree Plantations (Unit: 10^3 ha)

Year	*Oil trees*	*Fiber and pulp trees*	*Fruit trees*	*Total*
1962	89	813	82	984
1972	246	831	88	1165
1979	308	861	72	1241
1990	364	789	274	1427

Source National Forest Resources Inventory by 1990

agricultural regions. By 1990, DPRK fuelwood plantations accounted for 196,000 ha of land. Deforestation continued, but at a very controlled speed. Virgin forests were harvested in a scientific manner, thus remaining as forests with much less stocking volume. As such, the forest areas used for timber production remained stable during this period.

Period as a Lone, Ideological Socialist Country

Eastern European socialism collapsed in 1990, but the DPRK did not. However, its forestry industry did collapse. Economic progress for the heavy-industry prior development strategy relied on cheap oil supply; the nation used around 1 million tons from Russia during the late 1980s. This cheap oil source was interrupted after the Soviet Union collapsed. The products produced in the DPRK were largely traded with former socialist countries in Eastern Europe. These were unlikely to be sold to other countries after 1990, leading to an extreme deficit in foreign-currency purchasing power in the international market.

In the early 1990s, the DPRK harvested more timber. This was traded to China for the foreign currency it required for energy products. Estimates indicate that approximately 6 million m^3 of logs were sold to China (the majority of these were traded illegally). However, it is difficult to obtain official data from Chinese customs. This timber trade involved many official government sectors and military forces. As weather conditions in the DPRK are extremely cold during winter, logging for fuelwood in populated areas dramatically increased to supply raw materials for cooking and heating. Due to the poor supply of fossil energy, city citizens altered their heating systems to replace fuelwood as their primary source of heat. The National Report to CBD (2010) DPRK indicates that the firewood used by rural families amounted to 6–8 m^3 a year (in total, more than 7–8 million m^3 annually in the DPRK). Green mountain forests were denuded on a large-scale in the south and central parts of the DPRK, which were highly populated areas.

The DPRK was severely damaged by flooding from 1995 to 1996. During this time, successive floods destroyed 1.3 million ha of agricultural lands. These damages were estimated to cost US$1.038 billion in the agricultural sector in 1995 and US$2.271 billion in the overall economic sector in 1996. A drought in 1995 also devastated approximately 466,000 ha of arable lands. Even after 20 years of striving to adapt to these conditions, the DPRK has still not recovered. Since 1996, foodstuffs, energy, and nuclear power have become hot political issues between the DPRK and major international players, including the United States, China, Russia, and Japan. The DPRK has struggled to gain understanding from the international community.

Forest degradation increased instances of forest fires, insects, and diseases. A pine beetle outbreak occurred during the mid-2000s, resulting in severe losses to some red pine forests. I visited pine forests undergoing severe beetle outbreaks in 2006, at which time I advised DPRK officials that these losses could be reduced by killing the insects using large labor forces. The next day was Friday, which was a working day for DPRK officials. The mountains containing pine trees around Pyongyang were crowded with people. I heard nothing from the DPRK on the pine beetle issue after 2007.

Forest fires occurred quite often, thus accelerating the clearing of the forest cover. A 2012 NASA satellite photo showed a number of forest fires clustered along the DPRK's east coast; many of these fires produced thick smoke. Local residents blamed the lack of available facilities for the severity of the fires, but were also happy to see them burn because the cleared land could be used for farming and firewood.

Famine has been a problem for the DPRK since 1995. Hungry citizens engaged in illegal slash-and-burn farming at the fringes of forest regions. Some people immigrated to inner forests for clear-cutting and resettlement in remote mountainous regions.

Common consensus is that a solution to these issues and other crises (including those dealing with food insecurity, energy shortages, and nuclear power) can be achieved through an integrated and collaborative approach involving the DPRK, ROK, and multilateral international agencies, including the FAO, UNDP, WFP, and UNEP, as well as bilateral agencies such as the SDC.

The DPRK has made afforestation a national priority. As such, there have been “months of general mobilization for land management” in spring and autumn, while masses of people have been encouraged to plant trees and manage their allotted forests (MoLEP, 2010).

With support from the ROK, the Central Nursery Center was established during the early 2000s. The DPRK's own capacity expanded to approximately 9,000 ha, with a growing capacity of 1 billion seedlings by 2013. Each province and county contain individual seedling farms. In 2006, I assisted the DPRK in introducing a sort of electronic full-time irrigation system that was specialized so that cuttings could produce seedlings. This system was tested at the Central Nursery Center. This sort of facility was installed at every nursery farm at the county level after only a couple of years. However, I discovered that many technical issues impeded operations, including improper species selection, substandard cutting preparation, and poor machine quality.

Given the difficult situation in the DPRK, many international agencies and experts, including myself, considered that agroforestry was the last hope for saving the DPRK's forests. Since 2000, agroforestry has shown great potential through piloting projects financed by many multilateral and bilateral agencies. This practice was officially adopted by the DPRK in 2013. A central non-standing agroforestry committee was formed in March of that year, while an Agroforestry Regulation that was adopted by the DPRK Cabinet Resolution No. 35 on May 12. Article 12 states that “forest land use licenses” can be issued to user groups and individual households. License holders can legally clear sloped forest lands for agroforestry practices. The motivations and interests of user groups and individual households can thus be ensured as the government commits to providing materials and technical support to maximize the output from these lands. However, there are obvious barriers to agroforestry development. The government has accepted economic incentives for individual farmers and farmer groups to plant trees on sloped farmlands that were once considered illegal. However, state-owned trees and crops cultivated on the same lands can be owned by individuals, thus reducing or negating the incentive for farmers to plant trees on sloped lands to promote the agroforestry system. My best estimates indicate that approximately 1.5 million ha of sloped forest lands have been totally denuded and illegally cultivated by individual farmers or farmer groups. Agroforestry is the best option for this sort of land management. The National Agroforestry Strategy and Action Plan was thus issued in 2014. This measure aims for 300,000 ha of tree-crop intercropping by the end of 2024. However, there are major challenges to be faced ahead for instance, different groups may have opposing views or it may be left undecided who will plant, manage, and benefit from these trees.

Understanding Deforestation Drivers

Over the course of one century, forest cover on the DPRK decreased during Japanese colonization, slightly increased after liberation, decreased again during the Korean War, subsequently increased due to economic progress in socialist countries, and then dramatically decreased after 1990, when Eastern European socialism collapsed. Considering the size of its territory and forest resources, the DPRK may be the world's worst country for deforestation and forest degradation. Forest acreage decreased from 8.1 million ha to 6.9 million ha between 1990 and 2010, while growing stock decreased from 526 million m^3 to 330 million m^3 (nearly 3% annually) during the same time.

Grassroot efforts have been blamed for this deforestation and forest degradation. For instance, large amounts of fuelwood were harvested for cooking and heating while slash-and-burn cultivation was used for food crops. It is usually the rural communities living in forests that are blamed for any associated problems. DPRK government officials blame the lack of seedlings, pesticides, and fertilizers (a result of international sanctions) for their failure to reverse these problems. From an international perspective, however, corruption and poor governance are responsible for the major levels of deforestation and forest degradation seen in the DPRK (Liu et al., 2016). The DPRK case is a black swan. That is, it comprises the same people with the same governance regime operating on the same land, but forest coverage varies considerably. It is not known how current international politics or domestic development strategies and governance regimes will affect these forest cover changes. As such, the DPRK may serve as a case study for documenting lessons and learning from experience, although it is extremely painful for the people who live on the Korean Peninsula. However, the Central Nursery Center was established with help from the ROK in the early 2000s; with the urgings of DPRK leaders, this center has expanded. Even so, I have the same question: is the lack of seedlings a root cause of DPRK deforestation and forest degradation? The answer is a firm no. Who should be blamed for producing this inaccurate discourse? I think the ROK may have at least contributed.

The lack of forest science and technology is another excuse for the DPRK's failure to restore its forests. The Central Nursery Center has demonstrated how advanced technology and facilities can produce seedlings. Agroforestry has also gained legitimacy; it is promoted as a land use innovation produced through advanced technology. However, agroforestry involves typical local or traditional knowledge. As such, slash-and-burn cultivation is considered an agroforestry practice that could promote sustainable land use through lengthy rotations (Luo et al., 2015).

Clearly, not all decentralization policies can produce effective and efficient forestry management (Ribot, 2005). Some produce outcomes similar to those of the centralization process. Forests are arenas of struggle and conflict in which both trees and local forest dwellers usually find themselves on the losing side (Doornbos et al., 2000; Liu, 2006). Illegal user-group farming on sloped land greatly contributed to food security in rural areas of the DPRK around the year 2000. These practices have also been used to improve recent food production rates, a circumstance that can be considered a product of unofficially decentralized land management. Land use licenses issued to user groups can legalize land use rights. However, a legal approach to centralized land use planning may cause land use conflicts between users and the government as well as between non-tradable and tradable productions. Government intervention in policy reforms can also activate "silent" conflicts that may create new problems among forest dwellers, between the people and the government, and between the poor and rich (Liu, 2006). This requires new policies and institutional arrangements at different levels of the government, local communities, and the private sector, as well as financial delivery mechanisms, appropriate extension services, and effective monitoring and evaluation practices (FAO, 2006; Liu & Innes, 2015).

Drivers of Deforestation and Forest Degradation

The DPRK regards nature merely as a means by which to increase the living standards of its people. Thus, its government believes that the forest can be exploited for this purpose and has actively tried to transform the natural landscape for its benefit. The primary goal of the DPRK's forestry policy was to secure a stable supply of forest products, such as timber, in order to construct a socialist society. In addition, the use and transformation of forests to resolve scarcity of natural resources and farmland was a crucial economic issue.

The main factors of deforestation and degradation are reckless extraction of timber, fuelwood, crop cultivation in slopping land, and natural disasters. Therefore, carefully analyzing the reasons for deforestation and degradation are important for reducing and preventing those risks and encouraging forest rehabilitation.

Logging

Timber is one of the worlds' main forest resources, and indiscriminate timber cutting activities has resulted in increased deforestation and degradation, which has caused severe environmental issues. Large-scale logging in the DPRK started during the Japanese occupation, which saw an introduction of large-scale clear-cutting and deforestation and degradation of natural forests.

In the 1960s, the DPRK introduced a cyclic felling policy to ensure the supply of timber was maintained in a sustainable way. This sought to restrict deforestation and forest degradation by timber extraction through a combination of forest management and timber production. By 1990, the DPRK had undergone a period of rapid economic development. The cyclic felling policy had been well implemented by this time, and forest quality had improved and quantity increased, while annual production from the timber forests remaining remained about 4,400,000 m^3.

However, maximizing timber production with so-called scientific approaches did not prevent deforestation in the DPRK. Forest resources have been excessively exploited by the state. After the Korean War, the DPRK moved toward "heavy industry preferential development," to rebuild itself and create a socialist economic system. The forestry policy mainly focused on timber production. Timber production increased from 2.3 million m^3 in 1956 to 2.9 million m^3 in 1961 and has increased continuously since then (Park et al., 2013). As the DPRK pursued timber production, the forests were no longer protected. By the mid-1990s, the DPRK's economic situation had worsened, and it had faced continuous natural disasters, which meant local forest protection roles assigned by the central government were misused and local governments were responsible for forest despoliation. Deforestation and degradation caused by over extraction of timber brought about severe water and soil erosion control and caused issues in flood management and biodiversity conservation.

MoLEP attempted to reset the cyclic extracting, and the DPRK reduced the quota of timber production per year to 2,730,000 m^3. This total amount was then allocated to various organizations, including schools, hospitals, factories, mines, and railways, including any construction projects approved by various authorities. In particular, the DPRK reduced a large amount of timber supply to boat manufactures, then reduced production of fishing boats and fleets, and then reduced the production capacity of sea fishing.

Reclamation of Forestland for Farming

As food consumption rapidly increased after the late 1960s due to population growth and economic development, the DPRK undertook mass reclamation of forest land with terracings and encouraged people to cultivate terraced fields. The reclamation of forest lands for terraced fields was regarded as nature remodeling and was one of five policies for nature remodeling in the mid-1970s. A forested area of 0.2 million ha with a slope of less than 16 degrees was cleared and transformed into farmland (Park et al., 2013). This large-scale cultivation of terraced fields in forest areas led to forest degradation and caused floods and landslides in the western part of the DPRK. By the mid-1990s, the DPRK prohibited the excessive conversion of forests.

Fuelwood Extraction

Wood has traditionally been used as fuel for cooking or heating, and its usage has increased with population growth. After being replaced by coal, patrol, gas, electricity, and biogas, fuelwood consumption reduced in the 1970s and 1980s. As the DPRK faced chronic energy shortages since 1990, forests became the main household energy source. People started cutting down trees to obtain firewood for cooking and heating; the amount of cut firewood increased from 3 million m^3 in 1990 to 7.2 million m^3 in 1996. In the late 1990s, fuelwood consumption increased to 7,000,000~8,000,000 m^3 a year, which was a major cause of forest degradation in the DPRK. According to a census carried out by the Unite Nations Population Fund (UNFPA) in 2008, timber was the main energy source for heating and cooking, proving that the DPRK has a high dependence on forest products, especially in the countryside (UNFPA, 2008).

The DPRK took some positive steps in preventing deforestation and degradation of forests from over collection of fuelwood. In 2010, 100 ha of state forests were allocated to every cooperative farm (Ri) for fuelwood collection, and natural forests with high regeneration abilities and high productivity were selected as fuel forests for farmer cooperatives. This was to encourage the development of fast-growing tree species of high caloric value and to actively introduce ultra-anthracite coal and biofuel as alternatives or renewable energy together with fuel-saving stove.

Extraction of Non-timber Resources

The DPRK is abundant in high valuable non-timber resources such as oil-bearing fruits, so the residents in mountainous areas have used them for their subsistence and income generation. These have been used as raw materials for local industrial factories where food produce and traditional medicines rely on non-timber resources. Distribution of the typical non-timber resources is as follow (see Table 7.2).

The ratio of chronological extracted non-timber resources is as follows (see Table 7.3).

Table 7.3 shows that the amount of extracted wild fruit increased, but the increase in extracted wild vegetables and medicinal herbs was only gradually. The increase between 1979 and 2010 does not represent an absolute value of resources, but shows that the ratio is increasing due to a reduction in the amount of medicinal herb, wild vegetable, and mushroom resources among total resources.

The continually rising demand for natural foods and traditional medicines was caused by over extraction of non-timber forest production

Table 7.2 Selected species for non-timber forest products

Wild fruits	*Medicinal herbs*	*Wild vegetables and mushroom*
Chestnut, acorn, wild grape, blueberry, *Actinidia arguta*	*Schizandra chinensis*, *Angelica gigas*, *Epimedium Koreanum*, *Tractylis ovata*	Bracken, bellflower, *Aralia elata*, anise, *Tricholoma matsutake* (pine mushroom)

Table 7.3 Ratio change in amount of extracted non-timber forest products

Year	*Wild fruits*	*Medical herbs*	*Wild vegetable or mushrooms*
1979	90.2	5.6	4.2
1990	91.7	4.6	3.7
2000	93.8	3.4	2.8
2010	99	0.3	0.7

as well as demand for medicinal herbs, wild vegetables, and mushrooms that were used as raw materials for traditional medicine and healthy food.

Natural Disasters

Due to the steep mountains and upstream slopes in the DPRK, mountain erosion such as landslides are common. In addition to a decrease in the fertility of land resources, debris flow has also caused land loss and burial, and directly reduced land area. Deforestation and forest degradation have led to water and soil erosion, a decline in soil fertility, increased gully development, reservoir siltation, river blockages, and a rise in river bed levels, creating great loss of fertile farming land along river courses, and loss of life and property due to uncontrolled floods. Landslides and soil erosion are two of the main causes of deforestation and forest degradation in the DPRK. For example, about 80 spots in mountainous areas were recorded as devastated by landslides, including in Yangdok County, Songchon County, and Sinyang County, where 200 ha of land was buried or washed away in 2010. In 2012, over 140 spots were damaged by landslides and about 340 ha of land was buried or washed away in Songchon County. In that year, 43 spots were damaged by landslides, 200 ha of forest land was severely devastated, and 140 watercourses were wiped way in Unsan County (*Source* Central Statistic Bureau, 2008).

The DPRK has realized that it is a severe issue and has begun conducting forest management led by the watershed unit, with the aim of raising the forest plant coverage rate as much as possible to prevent deforestation by flooding. Afforestation for erosion control is being carried out in vulnerable regions to areas affected by landslides and erosion, and more areas, including reservoir forests, have been given protected forest status in an attempt to prevent flooding.

Forest Fires

Fires are a root cause of deforestation and forest degradation in the DPRK. According to yearly fire reports, the number of outbreaks and damaged areas has rapidly increased. Between 2001 and 2005, over a hundred forest fires had been recorded, with 5,000~8,000 ha listed as damaged areas. In 2009, this area had increased to 29,760 ha, covering a number of provinces, including: North Hamgyong Province, Jagang Province, and Ryanggang Province, which combined accounted for over 50% of damaged areas, North and South Pyongan Province, Gangwon Province, and North Hwanghae Province, which accounted for 35% and South Hwanghae Province Pyongyang City and Nampho City, which accounted for 15%. These results can be explained by the fact that food producing activities are centered mainly in central and northern mountainous areas, rather than plain areas. Combustible pure forests such as pine trees are widely distributed, which should also be considered.

The frequency of fires in western regions has increased, which is surprising as there were previously no forest fires in this part of the country. The DPRK has made a plan for the development and introduction of forest management methodologies like fire-free tree plantation. Forest officers lack fire management training and technologies. Local forest officers often complain of a lack of available equipment to put out fires. However, lack of community support could also be a major challenge. According to western media reports, residents are content with these fires, as land is cleared for farming, and firewood will be readily available for their use.

"Many believe that these are gifts from above," said the report.

Forest Pests

Forest pests have been recognized as one of the dominant factors contributing to deforestation and forest degradation. In accordance with the tendency of break and spread of pests in the country, some confirmed species of pest have explosively broken out and caused damage to the wide forest areas. The table below shows the ratio of areas damaged by forest pests from 1954 to 1990 (see Table 7.4). As noted in the table, *Dendrolimus spectabili* accounted for over 90% of pest damage.

According to outbreak variation of the Dendrolimus spectabili caterpillar, an explosive outbreak occurred in the 1960s, and gradually eased

Table 7.4 Ratio of damaged areas by main pests

Total area of forests (%)	*Pest Species*			
	Dendrolimus spectabili	Cecidomyiidae *brachyntera*	*Acantholyda* Sp	*Lerya purchasi* Maskell
100	94.2	2.8	0.1	2.9

from 1984 to 2005. The outbreak status was classified as low level, with a value between 0.1 and 0.3% in the 1970s and 1980s, but another outbreak occurred in the early 2000s. Damaged forest area exceeded 0.4% of total healthy forest area in 2006 and peaked in 2009. Damaged areas were each affected by different species.

Dendrolimus spectabili: Outbreak began in 2001 and peaked in 2007. In the west sea coastline, pine forests were nearly diminished by 2010. The area of damaged forests in the western lowlands worked out at over 90% of total area. In other regions, the percentage of pine forests destroyed were as follows: Pyongyang City 5.7%, South Pyongan Province 13.2%, North Pyongan Province 38.4%, South Hwanghae Province 20.2%, and North Hwanghae Province 11.4% including Kaesong City 3.3%.

Cecidomyiidae brachyntera: The forests in Gangwon Province and Kaesong City were most severely damaged by Cecidomyiidae brachyntera, but the damaged area has gradually expanded across the country (Gangwon Province 60.1%, Kaesong City 22.5%, North Hwanghae Province 9.2%, Pyongyang City 3.3%, and South Hwanghae Province 4.0%).

Lerya purchasi Maskell: This pest crossed over from China and is extremely destructive, although mainly present in North Pyongan Province.

Acantholyda Sp: Forests damaged by Acantholyda Sp were largely focused in the middle region of North Pyongan Province, which accounted for 28.7% of total damage (North Hwanghae Province 45.9%, Gangwon Province 18.0%). This pest is expanding its occupancy in the central regions.

Due to the geographical distribution and concentration of the main forest pests, rational measures that correspond to the characteristics of each specific area is necessary for quarantine, preliminary observation and prevention. However, pest control methods and forest management focused on prevention and mitigation are yet to be developed and implemented. Officials also have an outdated understanding of appropriate forest management approaches. The DPRK intends to introduce biological and managerial control methods to prevent further outbreaks and spread of forest pests.

References

Central Statistic Bureau. (2008). *Statistical report of population in DPRK.*

Doornbos, M., Saith, A., & White, B. (2000). *Forests: Nature, people and power* (pp. 1–3). Blackwell Publishers.

FAO. (2006). *Global forest resources assessment 2005* (FAO Forestry Paper No. 147, pp. 28–30).

Liu, J. (2006). *Forests in the mist—Livelihoods and responses to the natural forest protection program in China* (Wageningen PhD Dissertation, p. 87).

Liu, J. (2009). Reconstructing the history of forestry in Northwestern China 1949–1998. *Global Environment, 4*, 188–219.

Liu, J., & Innes, J. (2015). Participatory forest management in China: Key challenges and ways forward. *International Forestry Review, 17*(4), 477–484.

Liu, J., Liang, M., Li, L., Long, H., & De Jong, W. (2016). Comparative study of the forest transition pathways of nine Asia-Pacific countries. *Forest Policy and Economics, 76*, 25–34. https://doi.org/10.1016/j.forpol.2016.03.007

Liu, L., Liu, M., Xu, Y., & Liu, J. (2015). Changes and tendency analysis of forest land area in the key stated-owned forest region: A case study based on the data of forest industry region of Heilongjiang province. *Scientia Silvae Sinicae, 52*(8), 88–94.

Luo, Y., Liu, J., Zhang, D., & Dong, J. (2015). Actor, customary regulation and case study of collective forest tenure reform intervention in China. *Small-Scale Forestry, 14*(2), 155–169. https://doi.org/10.1007/s11842-014-9279-1

MoLEP (Ministry of Land and Environmental Protection). (2010). *Report of national forest resources inventory.*

Park, Kyung-Seok, Lee, S.-Y., & Park, S.-Y. (2013). Changes in The DPRK's forestry policy. *The DPRKn Review, 9*(1), 20–29.

Ribot, J. C. (2005). Choosing representation: Institutions and powers for decentralized natural resources management. In J. Carol, P. Colfer, & D. Capistrano (Eds.), *The politics of decentralization* (pp. 86–106). Earthscan.

United Nations Population Fund (UNFPA). (2008). *DPRK population census national report.*

CHAPTER 8

Afforestation and Reforestation

Afforestation and reforestation are regarded as the main approach for forest rehabilitation in the DPRK. The government is the most important actor in afforestation and forest restoration and plays a major role in planning, campaigning, resource mobilization, implementation, and monitoring.

Forest Landscape Restoration Internationally

Background and Concept

In the last few decades, landscapes worldwide have changed rapidly due to socioeconomic development and globalization, or in some cases (like the DPRK), due to increases in economic hardship and poverty. Diverse ecological and cultural landscapes are disappearing at a rapid pace around the world. Global forest degradation and fragmentation have led to a series of problems such as declining forest productivity, fragmentation of forest habitats, accelerated extinction of native tree species, environmental degradation, rural poverty, social conflict, and erosion of local cultures (Antuma et al., 2014; Aronson & Alexander, 2013: 4; Besseau et al., 2018).

We know that degradation can be reversed, and over recent decades many landscapes around the world have successfully been restored. Such restoration has often brought prosperity for humans and societies, and the

L. Jinlong, *Crying Forests*,
https://doi.org/10.1007/978-981-16-9725-8_8

restoration processes are often complex, providing economic, social, and ecological challenges, which require smart governance processes to overcome competing claims and conflicts over areas targeted for restoration and methods to be used (Van Oosten et al., 2021).

Restoration of degraded ecosystems is not easy. Many examples, such as the DPRK in the 1990s, have shown that even for simple afforestation activities it can be difficult to yield significant results. Certain forest rehabilitation projects only account for socioeconomic needs without account for long-term ecological consequences, or emphasize absolute protection, ignoring the diverse needs of different stakeholders (Erbaugh & Oldekop, 2018). Rapid expansion of plantation forests has been criticized for generating limited environmental services and, in the case of large-scale government-led forest restoration programs, for their failure to incorporate multiple stakeholders (Chazdon, 2008; Sayer et al., 2004; Xu et al., 2012).

In response to these issues, countries around the world are striving to implement a wide range of ecological restoration activities and going to great lengths to manage and preserve forest integrity. The causes of forest degradation are diverse, complex, and site specific, and so comprehensive approaches are required to achieve forest landscape restoration. Therefore, a more comprehensive approach to recovery should focus on the quality and quantity of forests while accounting for broader environmental, social, and economic needs, and the need to restore the ecological integrity of forests while also providing tangible benefits to local populations. Forest landscape restoration is emerging as an effective approach.

Based on the summary and analysis of a large number of forest restoration cases that took place in 1999, the IUCN, WWF, and other international organizations put forward the concept of forest landscape restoration. In 2001, forest restoration experts in Serbia and Spain proposed the concept of "forest landscape restoration," namely, the process of promoting human well-being while striving to restore the ecological integrity of degraded landscapes. Forest landscape restoration is the best way of reestablishing an optimal balance between ecological, social, cultural, and economic benefits of forests and trees under a multi-land use model, which is agreed among locals. In essence, restoration of forest landscapes is a dynamic, complex, and interactive approach to managing natural resources and land use, through coordinating land use between stakeholders. It is aimed at restoring ecological integrity and promoting the economic development of local communities and the country as a whole. The restoration of forest landscapes not only reflects the importance of forests, biodiversity, climate change, soil erosion, and desertification, but also contributes to the Bonn Challenge, which seeks to restore 150 million hectares of degraded or degraded land by 2020.

Forest and Landscape Restoration emerged in 2000 as a novel approach to regain ecological functionality and strengthen human well-being in deforested and degraded areas to address the negative impacts associated with forest restoration programs (Sabogal et al., 2015). The FLR approach expanded from ecological restoration and from reflection upon failures in conservation and forest management approaches and addresses interventions to recover or conserve native ecosystems. These interventions include farming and other initiatives to improve outcomes for local livelihoods, ecosystem services (ES), and biodiversity conservation at the landscape scale (Lamb et al., 2005; Mansourian et al., 2005). Because of the ecosystem services that forests provide, it is now widely acknowledged that forest restoration can contribute to goals pursued by sustainable development actors and related international initiatives. For instance, forests are expected to contribute to achieving the Sustainable Development Goals (SDGs) (De Jong & Galloway, 2020; Katila et al., 2019; Seymour & Busch, 2017; SOFO, 2018). They are expected to increase food and water security, boast rural employment and help protect biodiversity (SOFO, 2018). Forest conservation and restoration are key strategies to achieve the UNFCCC Paris Agreement as well as SDG 13 (Bastin et al., 2019; Long et al., 2018).

International Practice and Experience of Forest Landscape Restoration

Compared with traditional forest restoration, forest landscape restoration has the following characteristics: Firstly, it is based on landscape scale. Forest landscape restoration is not limited to a single site, but may be difficult to achieve and contradict land use needs in order to achieve unity in the landscape level. Secondly, the choice of technical and economic measures is constrained by human well-being and the sustainability of ecosystems. Thirdly, it is a comprehensive management process that values cooperation of relevant interest groups. A cooperative and coordinated political atmosphere is required, and a bottoms-up institutional system is needed that promotes consultation, collaboration, and participation from different stakeholders in the decision-making process, including individuals, enterprises, government departments, scientific research institutions, and social organizations. Fourthly, it is used to promote the balance of different social ecosystems and restore the holistic, health, and vitality of the landscape level. The holistic nature of the landscape level covers two or more interconnected ecosystems and is composed of different social ecosystems, such as virgin forests, plantations, farmlands, lakes, rivers, and villages.

Due to the work of international organizations such as the IUCN, WWF, ITTO, and FAO, many countries and regions are taking active measures to carry out FLR work. Before FLR, Gilmour (2005) explained the reconstruction of degraded forest landscapes in Vietnam, Laos, Cambodia, and Thailand through utilizing forest resource status, current forestry policies, and process introductions and analyzed experiences and lessons learned. After a definition of FLR was put forward, the African Regional Symposium (organized by the IUCN and WWF) summarized the findings of FLR in Tanzania, Uganda, Ethiopia, and Kenya. Maginnis et al. (2002) cited examples of forest restoration in Tanzania and Malaysia in analyzing the causes of forest restoration in the tropics; Lamb D et al. (2005) provides detailed information on the use of the FLR approach for the rehabilitation and restoration of degraded forests, and a detailed overview of the regional and global approaches to forest degradation in forest restoration in Canada, Nepal, Brazil, and New Zealand, as well as the implementation of FLR in central and northern Europe, western Finland, central Russia, and Scotland.

Since April 2005, seminars on the implementation of forest landscape restoration have been held in Petropoli, Brazil, and FLR practice and research have undergone rapid development as a result. The postponed Declaration on the Petropoulos lists successful case studies of countries that have implemented the FLR approach to improve important ecosystems in degraded or deforested forests, including Tanzania, the United Kingdom, Brazil, China, India, and Mali among other countries. The IUCN and WWF published the Arborvitae, which detailed examples of forest restoration, including the restoration of the Tanzanian Ngitili (a forest that plays a vital role in local livelihood), mangrove wetlands in India, ancient woodlands in Britain, as well as the restoration activities supported by the WWF, including projects to restore degraded watershed near a Peruvian temple in 2004 and restoration of the cork oak forest landscape in Portugal.

Lessons Learned from International Practice

Local involvement networks should be involved in all aspects of the project life-cycle, from planning, implementation, and management, and active involvement of key stakeholders, including locals, is essential. Agroforestry policies should focus on the development and strengthening of local organizations. There are many tools and extensive literature on local participation and cooperation in forest landscape restoration. The main contributors include Borrini-Feyerabend, the FAO Community Forest Management Series, and training materials provided by the Asia Pacific Community Forestry Training Center in Bangkok—these include training materials that cover social communication, information gathering and assessment, local organizational development, planning, implementation, local knowledge, conflict management, and participatory tools and processes for monitoring and evaluation. The Center for International Forestry Research (CIFOR) has developed a range of interactive tools for collaborative learning, community participation and organization, conflict management, and also community-managed landscapes, plantation landscapes, and rehabilitation. The use of local knowledge in landscapes has standards, indicators, and guidelines. A number of tools have also been devised by local forest dwellers to promote the use of local knowledge in the development of sustainable forest management criteria and indicators.

The chosen methodologies should address the socioeconomic needs of local people, thereby ensuring their continued involvement and interest in projects and improving their livelihoods. In some cases, reconstruction projects have actually deprived locals of their livelihoods (such as by restricting agricultural farming on redevelopment areas). In the Philippines and Vietnam, for example, project beneficiaries often cause fire damage to project areas, so that they can be reemployed to plant trees or rebuild forests. Therefore, there is an urgent need for socioeconomic analysis of future production systems and small-scale surveys to determine whether local farmers and communities can benefit directly from reforested forests before promoting forest rehabilitation. The DFID Sustainability Livelihoods Toolkit provides tools for increasing sustainable living in different phases of the project cycle, from planning and implementation, to monitoring and assessment. The ITTO Landscape

Restoration Guide offers a variety of recommendations on livelihood-enhancing activities, including assessing the prospects for forest products and payments for eco-environmental services, and assessing different land use patterns. CIFOR and other agencies have developed indicators to measure and assess the socioeconomic impacts of different projects, processes, and policy changes.

Clear and appropriate institutional support and appropriate institutional arrangements are essential to facilitate investment and local participation in reconstruction projects and ensure project sustainability, including very clear and uncontested land tenure. Equally important are formal institutional arrangements that include a comprehensive and participatory process in which participants, rights, costs, and benefits are clearly defined. Clear institutional arrangements can help prevent conflict, promote coordination, and fulfill established mandates, ensuring that different stakeholders have access to mutually agreed benefits, and fully back long-term success of projects. CIFOR sets general indicators for institutional arrangements, land ownership, and legal frameworks to ensure that large-scale plantation landscapes managed by communities are sustainable. Implementation of agreements is an important part of this institutional arrangement.

Many degraded forests occur due to various land uses, requiring governance across sectors that account for the needs and aspirations of multiple stakeholders within the landscape (Chazdon et al., 2016; Sayer et al., 2013). Forest landscape restoration moves restoration beyond sectoral and site-based activities toward an integrated approach at landscape level (Hobbs et al., 2014; Newton et al., 2012; Ros-Tonen et al., 2014). The adoption of forest landscape restoration requires a significant shift from state-controlled to polycentric governance, which is characterized by multiple governing authorities at multiple levels (Ostrom, 2010). On the one hand, the failure of large-scale government-led forest restoration programs has prompted the innovation of forest landscape restoration both theoretically and practically (Clement, 2010; Mansourian et al., 2005). On the other hand, globalization, decentralization, and marketization trends have increased the diversity of actors and the extent and depth of participation, shaping forest governance at the local, regional, national, and global levels (Agrawal et al., 2008). This introduces new actors, institutions, networks, knowledge, and instruments that reshape the relations of power and cooperation in forested landscapes (Görg, 2007; Monroe &

Butler, 2016) and transform forest landscape restoration toward polycentric governance (Guariguata & Brancalion, 2014; Mansourian & Vallauri, 2014; Van Oosten et al., 2014).

China's Experience

China has been particularly successful in addressing restoration pitfalls. Historically, the country has been confronted with severe landscape degradation due to massive political, socioeconomic transformation leading to large-scale deforestation, water and soil erosion, and loss of biodiversity. However, rapid response actions have led to massive restoration of China's degraded lands. A frequently mentioned example is the northwestern Loess Plateau, which was highly degraded, but completely restored within a decade. Similarly, the country has been able to successfully reforest and restore landscapes. For example, between 1998 and 2013, China has restored about 3.6 million hectares of forests annually, which equals about 35% of its total global forest loss. This enormous effort has been made possible by strong government interventions and massive investments in the dryer hinterland areas, as well as in the major watersheds surrounding China's coastal cities. In this way, China has managed to restore large parts of its productive land, while generating employment and providing ecosystem services to metropolitan areas nearby.

From Grand Design to Co-design

Large-scale state intervention, drastic top-down planning, and paid labor programs were the engine behind massive landscape transformations in the past. However, more recently China has developed a more participatory and "bottom-up" approach to restoration. New forms of hybrid governance and public–private collaboration have allowed for more integrated approaches involving different stakeholder groups, including local governments, NGOs, environmental organizations, landowners, companies, and investors. This cross-sectoral strategy has brought new dynamics to China's rural and peri-urban landscape.

It is essential that the participation of multiple actors and the application of market-based instruments facilitated a transition from state-centered to polycentric governance in forest landscape restoration. The Conversion of Cropland to Forest Programme (CCFP) and the Sloping

Land Conversion Programme (SLCP) illustrate this transition. The CCFP (1990–2000) had a strong focus on regional diversification and local governance. Subsidies were provided to farmers to "retire" their croplands and convert them into forestland, which led to the largest conversion of cropland to forestland the world has ever seen. Cross-sector networks were created, to allow for horizontal cooperation at the local level. Combined with forest tenure reform and new digital banking technologies, CCFP allowed for rural modernization and financial inclusion of rural households across the country as demonstrated in the research by Kun Zang and colleagues.

The SLCP was launched in 2001 and continued to subsidize farmers for "retiring" and restoring degraded croplands. Decision-making was further devolved to the local level, and vertical networks were created to strengthen local democratic institutions and enable village administrations to better control their own resources for sustainable use, according to researcher Jun He. Although the impact differs from place to place, the promotion of local autonomy has had a positive impact.

Innovative financial instruments provided monetary incentives for farmers to diversify their livelihood strategies; crop yields were enhanced, farm sizes reduced, and forest areas enlarged.

New Roles of Local Actors in the Public and Private Sector

Apart from these large-scale national programs, there has also been a growing number of local level projects, such as the restoration of the Miyun watershed. As described by Van Ham and colleagues, the Miyun watershed is Beijing's main water source. Initiated by the city's government, this ambitious project has been implemented in close collaboration with rural (upstream) and urban (downstream) inhabitants, water dependent private companies, international NGOs, and the municipality itself. The basic principle of the project is that water users (companies and urban citizens) pay for the restoration of upland forests, while inhabitants of the uplands undertake the work. Restoration is implemented through a range of practical options, which serve ecological purposes but also provide alternative livelihood options for rural dwellers, such as fruit trees, timber trees, apiculture, horticulture, and other remunerative activities. A multifunctional restoration model was developed, which is currently being implemented in many of China's megacities that are suffering from increased water stress.

New Capabilities Needed

So far, little is known about the practical implementation of this new national park policy, but it would be an excellent opportunity for further research, to see how win–win options can be created without compromising the livelihood of local people. Challenges may arise from the complex administration of large eco-zones stretching across different jurisdictions and covering multiple land uses with the risk of creating competing interests and conflict over space.

This will require park management and staff in areas of conflict mitigation, stakeholder collaboration, and policy integration—there is a chance that these resources are not currently included in the curricula of natural resources management education. Increased management and staff capacity may be urgently needed to cope with new challenges. Also, a review of the curricula on forestry and natural resource management education institutes may need to be revised and developed in order to cover future national park management and staff.

Forest Landscape Restoration in the DPRK

Forest Landscape Restoration Initiatives in the DPRK

The DPRK is mainly comprised of forest land (80% of the total land area), but arable land is very limited, amounting to only 20%, and located mainly on elevated land. The land is also susceptible to soil erosion. Forest resources have a significant impact on local livelihoods, economic development, social stability, and ecological security, and so forest degradation and destruction have naturally resulted in poverty and climate change, and caused issues with food safety and biodiversity—some of the main problems facing the DPRK today (Jin et al., 2016). According to FAO (FAO, 2010), the DPRK's forest area reduced from 8.2 hectares to 5.67 hectares between 1990 and 2010, and so average forest area amounted to 5.67 hectares during this period. Forest volume also decreased from 5.26 billion cubic meters to 330 million cubic meters per year. Data from the DPRK's Ministry of Land and Resources (MoLEP, 2011) shows that 4.54% of plants and 21% of animals face endangerment.

In practice, the DPRK has carried out a number of forest rehabilitation activities. However, due to unavailability of reliable data, only a small amount of research has been conducted by the academic community on the restoration of Korean forests, and so there is no consensus or complete description of the country's forest landscape restoration. Prior to the 1990s, there was a period of strong Sino-DPRK relations. Experts from the Ministry of Forestry in China described the restoration process of forest landscapes in the DPRK in a report, which stated: from 1945 to 1950, DPRK forests were owned by the government, and forest management orders were issued to mobilize civilians to plant trees. War occurred between 1950 and 1953. During the period of full recovery and development from 1954 to 1980, nationwide reforestation was implemented, and a tenfold and circular harvesting policy was adopted. After 1981, the country entered a period of intensive management and restored 2 million hectares of forest. By 1983, nearly half of DPRK forests had been nurtured after liberation, with 75% forest coverage (Li Shixian, 1985). Since the 1980s, Chinese experts were unable to visit the DPRK due to weakening Sino-DPRK relations. The only available information on forest landscapes after the twenty-first century in the DPRK is a translation of Professor Kim Il Sung from the Northeast Asian Studies Center of Jilin University. Report: After the twenty-first century, the DPRK put forward the goal of "National Forestation and Greening." The DPRK formulated an annual plan through scientifically examining characteristics and natural conditions of mountain forests. The plan involved planting high value forests, with species such as the Japanese larch, Populus alba, and Ginkgo biloba. The plan also sought to strengthen forest fire prevention measures, expand forest areas, and use forest resources to improve living standards and promote economic development (Jin Zhe, 2016). Official readings by the Korean forestry authorities detail the complete restoration process of forest landscapes since the founding of the People's Republic of Korea. The process was divided into two phases: The first stage was the afforestation phase, which took place from 1946 to 1960. It had the goal of rehabilitating 850,000 hectares of forest land and establishing a forest plantation (totaling 239,000 hectares) in two years to achieve afforestation and soil erosion management. The second stage was the biodiversity conservation phase, which took place between 1961 and 2013. The aim of this phase was to restore 850,000 hectares of forest land and to establish a commercial forest-based system to ensure timber production. It remains to be determined whether this reporting on the restoration process of forest landscapes by the DPRK Forestry Ministry was purely part of political propaganda.

Degradation and destruction of forests in the DPRK have attracted widespread attention from the international community (Laurance et al., 2014; West et al., 2014). The DPRK is connected to the mainland and the sea, and as it is home to a diverse range of biodiversity, conservation is particularly important. Domestic afforestation has been insufficient in curbing forest degradation and destruction trends, and so FAO and other international non-governmental organizations introduced the Forest Landscape Restoration Project in the DPRK; the DPRK was chosen to be one of nine pilot countries for the 2014 Forest Landscape Restoration Project, which sought to assist the DPRK in restoring 1.5 million hectares of forest. Since the 1980s, decentralization reform and international environmental politicization have given rise to a large number of non-governmental actors, such as NGOs, multinational corporations, communities, forestry farmers, and consumers. The state's influence and power have been consistently challenged (Agrawal et al., 2008). Forest governance has attracted attention from a growing number of actors outside of government, such as multinational corporations, international environmental NGOs, research institutions, indigenous people, consumers, smallholder farmers, and the forest product industry. Developing countries have also integrated into the international environmental governance system. As a result, forest landscape restoration strategies in the DPRK have changed, as have attitudes toward external support. FAO's Forest Landscape Restoration Project was successfully implemented in the DPRK after 2014.

Afforestation Plan

Green DPRK has been on the political agenda for many years, as forestation was promoted as a way of boosting the national economy and improving living standards. The DPRK made amendments to the Master Plan for Forestation periodically (see Table 8.1), which guided national forest development. Table 8.1 provides a summary of afforestation targets since implementation in 1945.

As noted in Table 8.2, area of plantation in 1990 was 4.1 times that of 1957. Consequently, non-wooded forest land accounted for 6.2% of total forest area in the early 1990s.

It is extremely hard to obtain data on the progress of master afforestation planning. There were undoubtedly achievements pre-1990, as is

Table 8.1 Afforestation plan and output since 1945 (Unit: 1000 ha)

Year	*Plan*	*Output*	*Duration*
1946	–	36	–
1948	56	–	1-Year Plan
1949–1950	239	89 (1949)	2-Year Plan
1953	–	100	–
1954–1956	189	52 (1956)	3-Year Plan
1957–1960	500	508 (1957–60)	5-Year Plan
1961–1967	850	914 (1961–70)	First 7-Year Plan
1971–1976	–	48 (1972) 38 (1973)	6-Year Plan
1978–1984	510	100 (1975)	Second 7-Year Plan
1980	2,000	–	–
1987–1993	1,500	–	Third 7-Year Plan
2000–2010	1,500	1,000	Ten-Year Master Plan for Forestation
2015–2024	1,680	–	Ten-year Master Plan for forest restoration

Table 8.2 Variation in the ratio of artificial plantation area

Year	*1957*	*1972*	*1979*	*1990*	*2000*	*2010*
Ratio of plantation in total forest area (%)	4.8	13.5	16.2	18.8	16.2	13.5
Total area with forests (1,000 ha)	/	/	8435	8133	7554	6912

Source General forest resource inventory, 2010

evidenced by the increasing acreage of planted area during this time. However, the ratio of plantation in total forest area and areas with forests in the DPRK has continually decreased, despite 1 million ha of afforestation being implemented in the area during the Ten-year Master Plan for Forestation (2001–2010).

In 2014, the DPRK formulated a 40-year strategic forest restoration master plan. In this plan, the 10 years Forest Restoration Master Plan was devised, which aimed to afforest 1.68 million ha of land. DPRK officials in central government, provincial government, and officials in the county level People's Committee were interviewed and firmly believed that these targets could be achieved. However, responses from international experts who were interviewed disagreed and believed this target could only be achieved if the DPRK put 90% of its resources toward raising nurseries, and 10% toward managing forests. From what I witnessed out in the field, I believe it is highly unlikely that these goals will be achieved.

Methods for afforestation. The public are a major force used for planting. All Koreans classed as "able" are required to participate in the planting season, from students and public servants, to army officials, workers, and farmers. This rule has persisted to modern times. However, seedling technology has improved and mixture plantation introduced.

Guiding Principles for Afforestation. The DPRK prioritized afforestation for economic growth and paid less attention to environmental protection. For instance, in order to ensure forest land use was maximized, there was an increase in intensive use of forest land for planting thick and fast-growing trees with high commercial value. Additionally, afforested mixed forests were expanded to maximize productivity of land. In mountainous and plain areas, many forests of economic value such as metasequoia and walnut forests were afforested, as they contributed to national economic growth and improved livelihoods. Research on plants with high economic value was prioritized, and empirical findings resulted in extensive planting of those specific trees nationally.

Seedlings. Up until the early 1980s, only bare-root seedlings were produced for forestation. But from the 1980s onwards, container seedling technology was developed. Roll of vinyl sheetsor mesh potting was developed, and humus pot was later introduced by the end of 1990s. In 2006, with my assistance, MOLEP introduced a technology called the "Full-Sunshine Water Spray Seedlings System," which was rolled out so that each county has at least two systems to produce seedlings. At present, bare-root seedlings are used for seedling production. In the afforestation sector, nursery tree production and seed selection systems were established, which resolved many scientific problems including: premature ripening of pine nut trees, acclimation of bamboo, and standardization of afforestation technologies including distribution of tree-types based on afforestation group and plant thickness.

Afforestation model. Afforestation design follows three principles: (1) allocating land to increase land productivity (2) ensuring tree species are planted on appropriate land and (3) ensuring each tree species is planted in the correct season. Different types of afforestation include: whole planting, pitting up planting, and under-tree planting. Pure forestation practices gave room to mixed and multi-story forests, however I never came across any while at afforestation sites. Linear planting and belt planting methods have been introduced since the 1980s, and I had the opportunity to see this type of agroforestry development first-hand.

Plantation management. Based on official documents, technologies and operational standards are available, including thinning and tendering. Pest control methods that use bio-agricultural chemicals were reportedly developed and used extensively in the forest protection sector, but it was hard to assess the exact situation on the ground. Cultivation of edible and medicinal mushrooms has led to technological and scientific problems in the forest utilization sector.

Scientific and technological improvements. Technology is used in forestry to resolve issues related to the low survival rate of seedlings, and chemical materials are used to improve plant survival rates in drought season. There is limited interest in scientific and technological developments in this area, but further improvements are needed to understand why problems occur, as this will be key to developing approaches to improve seedling survival rates. There is also a lack of research into the social and economic factors that drive deforestation and forest degradation. However, technological innovation is a top priority for central governmental institutes, such as the Academy of Sciences and Academy of Forestry Research. There have been some efforts to promote the advancement of technology in forestry, for instance aerospace photographs, satellite images, and computer technologies have been used to monitor afforestation and forest management activities and have also been used to help establish seed selection and breeding systems. Forest management institutions have actively engaged in seed breeding conservation and management. In 2010, the production systems of three pilot stations—Huichon, Tongchon, and Songgan—underwent modernization, reconstruction, and expansion, which provided the foundation for mass production of hybrid young trees. However, there is no evidence to date that indicates final afforestation results.

Action Taken Between 2000 and 2013

Forests in the DPRK have continually degraded over time. However, countermeasures have been implemented in an attempt to revert this process. After 2000, the DPRK began forest rehabilitation and took steps to rehabilitate its degraded forests through its own efforts and in collaboration with South Korea and international partners. The DPRK set up the Ten-Year Afforestation/Reforestation Plan, from 2001 to 2010, and intended to plant a total of 1.5 million ha of forest area during its time, about 100,000–150,000 ha per year.

Firstly, nursery facilities and seedling capacity increased. The Central Nursery, managed by the Ministry of Land and Environment Protection (MoLEP), increased by 100 ha. In 2000, the Ministry of Land and Environment Protection (MoLEP) arranged for the restoration of destroyed nurseries and the establishment of a new central nursery with an area of 100 ha. In addition, the production capacity of this central nursery increased to as much as 400,000 nursery trees in the period 2006–2009. MoLEP organized the construction of mother nurseries that were 20–25 ha in size at municipal and county levels, which resulted in increased seedling production capacity in 2002—four times the production of 1994. From 2002 onwards, a nationwide campaign was implemented that saw the annual reforestation of over 900,000,000 trees.

Secondly, MoLEP promoted the creation of firewood forests with a total area of 800,000 ha to solve firewood shortages (the main cause of forest deterioration and soil degradation).

Thirdly, the DPRK government quadrupled the number of forest rangers in order to strengthen legal controls on illegal tree cutting and improve forest management.

Lastly, attempts have been made at proper harvesting. Deforested areas have seen an increase in active reforestation, which complies with cyclic felling principles. Similarly, areas of low productivity and severe soil erosion have been transformed by afforestation. Recommendations have been proposed that suggest concentrating felling to those areas with favorable cutting conditions and actively applying useful methods for lumber transportation.

The DPRK also enacted a policy that prevented further deforestation. In early 2009, the government prohibited tree felling and cultivation of forest land without permission, forcing people to sign "Contract for Forest Protection." Forcing people to sign this document was hard to justify and the consequences of signing were uncertain.

Provincial-level forest development design stations were responsible for establishing and implementing this ten year afforestation plan. The DPRK declared the plan a success, but there were no obvious improvements in the forestry situation.

Natural Rehabilitation

Natural rehabilitation will play an important role in forest ecosystem security and the protection of animal and plant resources in the DPRK. However, little attention has been paid to natural rehabilitation.

Generally speaking, natural rehabilitation has occurred in protected areas, as well as sparsely populated areas with limited human activities. Natural rehabilitation guarantees sustainable ecological security not only in forest reserves and northern highlands, but also in production-purpose forests. In the northern highlands, a lot of timber was produced as a result of rehabilitation. Also, many herbal resources from the naturally rehabilitated forests contributed to improved livelihoods and development of local industry.

Forest rehabilitation was unsuccessful in less densely populated areas of the DPRK due to over-harvesting of timber and fuelwood collection. Natural forests situated on sloping land were destroyed by cropping activities and over extraction of fuelwood. These types of activities are detrimental to the natural rehabilitation process.

The DPRK plans to restore almost 200,000 ha of forest land by natural rehabilitation between 2013 and 2023, in order to guarantee the security of forest ecosystems and prevent further deforestation and forest degradation due to activities such as overharvesting, slash-and burn cultivation, and overgrazing. The plan has a particular emphasis on forestation of wild fruit and fuelwood production. Guidelines for natural rehabilitation are currently being drafted, and scientific research in this field is increasing, with a focus on natural forest protection and management. I found no evidence to suggest implementation has actually begun at any of these sites.

10-Year National Forest Restoration Plan

The National Forest Development Perspective Strategy (2015–2044) was developed as part of the 10-year National Forest Restoration Plan (10-year NFRP). As part of this strategy, organizations, enterprises, and entities were allocated two months to plant trees and manage their allotted forests.

Goal

The main aim of the plan was to restore 1.68 million ha of deforested and degraded forest land. In order to achieve this goal, special attention was paid to the10-year NFRP through a series of afforestation and reforestation campaigns. This 10-year plan was divided into two phases.

The first phase, between 2015 and 2017, had the following goals:

(1) Plant trees in key sites such as highways, railways, villages, the Taedong River Basin, and boundary areas;
(2) Double the seedling production achieved in 2014;
(3) Decrease the timber dependence rate by 20% of the 2014 rate;
(4) Install fire lines and establish monitoring and early warning systems for forest fires;
(5) Establish and strengthen foundations such as productive bases for equipment and agrochemicals for pest management.

The second phase, between 2018 and 2024, has the following goals:

(1) Plant trees in all degraded and deforested areas including remote areas far away from communities;
(2) Produce more container seedlings—no less than 50% of the total annual seedling production;
(3) Decrease dependence on timber products to 50% of 2014 production;
(4) Maintain fire lines and complete afforestation for erosion control;
(5) Normalize the production of agrochemicals for pest management.

The following is a summary of key tasks in the first phase:

(1) Industrialization, intensification, and modernization of seedling production should be completed on nationwide scale. It should be widely introduced in every county forest management board.
(2) Globally recognized techniques should be adopted and tailored to promote afforestation and rehabilitation in the DPRK. The government planned to raise the socioeconomic and ecological value of forests by increasing forest fire prevention methods, increasing pest control, and protecting forest biodiversity through natural rehabilitation and mixed forestation.

(3) Local living standards should improve. Forestation and rehabilitation need to be increased rapidly, and agroforestry management should become general practice throughout the country. Guidelines and models are needed to modernize agroforestry.
(4) Forest biodiversity should be protected, which requires restoration of forest ecosystems. This can be realized by initiating the action plan on protected area management, and combining this with afforestation and natural rehabilitation.
(5) The scientific and technical problems associated with afforestation and rehabilitation should be solved by strengthening scientific research. It is necessary to increase scientific exchange between national and international scientific institutions.
(6) Bilateral and multilateral cooperation should be used to advance afforestation and rehabilitation methods.

Measures

According to planning documents, successful outcomes of this program include:

(1) Increasing afforestation and forest protection. Afforestation success includes: (1) Providing sufficient seedlings; (2) Planting trees in a responsible manner; (3) Cultivating trees in line with scientific and technical requirements. Implementation of the following measures would be deemed a success: (1) Prohibiting indiscriminative deforestation; (2) Efficient pest management; (3) Forest fire prevention; (4) Resolving firewood issues.
(2) Enhancement and application of forest science research and technologies. Measures involve (1) Tackling challenges associated with breeding, plantation, protection, and management; (2) Introduction and distribution of advanced technologies for plantation and protection; (3) Promoting the Academy of Forest Sciences. Plantation and forest protection should be strengthened, supervised, and controlled. Measures include: (1) Cohesive supervision; 2) Strengthening legal control of plantation and protection; (3) Distinction between good and poor units; (4) Enhanced capacities of forest rangers and increased awareness of their roles and responsibilities.

Progress to Date

According to official reports, progress by the end of 2015 was as follows:

(1) Forest restoration campaigns had been established from central level down to the county level. The General Bureau of Forestry and MoLEP administrated unified guidance on forests throughout the DPRK;
(2) Nurseries had been modernized and productive capacity of buildings had increased: The central nursery and all provincial nurseries had been updated; Seedling productive capacities of mother nurseries at city and county levels had been strengthened; there was increased capacity in nurseries/site nurseries managed by different organizations and farms;
(3) Improved plantation quality: (1) Most areas had been transformed into mixed forest; (2) Great improvement in the rate of rooting due to advanced methods; (3) Decentralized tree planting systems helped ensure high survival rates of planted trees.
(4) Forest fire control measures were implemented: (1) National forest fire prevention systems were established from central level down to provincial, city, and county levels; (2) National monitoring and alarm systems that used satellites were established; (3) There was an introduction of technologically advanced forest fire monitoring systems in certain regions, including Mt. Guwol; (4) There was an expansion of forest fire break lines in the most important areas.

Progress to date includes: (1) National institutional systems and forest restoration campaigns implemented from central government down to provincial, city, and county levels; (2) Extensive media and press attention; (3) Nurseries at the provincial and county levels have been updated (including the central nursery), and seedling productive capacity has increased; (4) Effectiveness of reforestation efforts has increased including increased survival rate of planted trees through enhanced tree planting; (5) National forest fire monitoring and alarm systems have been established, which seek to prevent forest fires and protect forests through the use of satellites.

Challenges

Many officials I came across in the DPRK had full confidence that the country could achieve its master plan targets, claiming that people of the DPRK could achieve great successes due to the country's strong leadership. However, those that worked in relevant agencies were aware of the key challenges faced by the DPRK in achieving its goals on forest and landscape restoration, namely:

(1) Insufficient awareness of landscape degradation and forest restoration.
(2) Insufficient experience in forest and landscapes, and limited technologies that could be used for implementing the plan.
(3) Weak local governance, such as inefficient forest management units and county level forest management boards.
(4) Strong need for external funding and mobilization of all resources, in order to successfully implement the forest restoration plan; Global efforts are also needed to promote good practice, as well as knowledge and experience sharing in forest and landscape restoration.

To do this, efforts are needed to increase tree seedling production and nursery capacity of high-quality tree planting materials. The NFRP states that 20% of these areas need to be restored through natural regeneration efforts. The plan has the intention of restoring deforested and degraded forest land not only through reforestation and afforestation but also through promoting agroforestry in sloping lands, and various protective measures including forest fire control and forest pest management.

All relevant bodies, from central to local level, have been involved in the implementation of the NFRP, namely: MoLEP, Ministry of Agriculture, Ministry of Timber industry, Ministry of Food Stuff Industry, and Ministry of Public Health. MoLEP played a key role in the planning and implementation of the NFRP, and the coordination of all relevant stakeholder ministries, institutions, and agencies.

Urban Greening

The DPRK considers eco-city construction as an important component for countrywide forestation and plantation. In the DPRK, the government laid down policy for civil gardening and greening in the past and established an institutional system for urban greening and research. For instance, gardening and greening affairs are managed by the Ministry of Municipal Management at a central level, by the bureaus of municipal management at the provincial level and by the urban management department and gardening boards at county level. There are also special groups that research urban greening in the Institution for Municipal Science. They carry out yearly trees, shrub and turf planting, and research and develop valuable tree species, seeds and planting methods for civil gardening.

Civil gardening is conducted in accordance with the urban greening plan, which includes provisions to prevent and reduce levels of environmental pollution, such as air and water pollution.

There is a plan to increase green area per person to more than 25 m^2 in urban areas. Plans have been formulated and are being carried out for creating shelterbelts along roads and near rivers, replacing vacant land in residential zones with green space, expanding green areas in parks and recreation sites, planting more flowers in urban areas, and increasing shelterbelts that separate residential areas and factories.

According to this plan, expansion of green track lands will be maximized to increase green track per head in urban regions. 6~10 trees are being planted for every 10 m^2 of road, a twofold increase. However, urbanization is causing challenges to civil greening. It's a shame that modern techniques have not been able to deal with issues of population growth and industrialization reducing green spaces.

Forests for Conservation and Protection

In 1998, the DPRK developed a National Biodiversity Strategy and Action Plan (NBSAP) at the request of the CBD Secretariat, with support from GEF/UNEP. The National Strategy and Action Plan included strategic issues and priorities in biodiversity conservation, as well as its measures and obligations in protection of the ecosystem. Article 6 of NBSAP was enacted in 1998 and included the following provisions:

1. Establishing a protected area system (including key areas), that will be significant in biodiversity conservation and management.
2. Restoring ecosystems damaged by natural disasters, and integrating biodiversity conservation plans into land use programs.
3. Bio-resource creation and establishing a system for its sustainable use.
4. Reinforcing laws and regulations on biodiversity conservation.
5. Intensify scientific research on biodiversity conservation.
6. Train experts and technicians in the field of biodiversity.

The DPRK has taken several measures with regards to the expansion of protected areas, improvement of watershed management, and the development of agricultural production, and established a national framework on bio-safety. However, diversity has worsened due to the decrease in forest resources, and ecosystems and the environment have deteriorated due to unsustainable use of biological resources. With the support of UNEP, the DPRK updated NBSAP in 2010. NBSAP was developed in consultation with various stakeholders, and attempts were made to identify threats and constraints in biodiversity conservation. Implementation mechanisms, strategies, plans, and action points have been designed so as to integrate biodiversity concerns into various other sectors. Attempts have been made to align the NBSAP with the ecological, social, cultural, and economic mosaic of the country and provide a focus and impetus to the current efforts toward biodiversity conservation. The updated NBSAP has reviewed and analyzed national activities regarding CBD implementation in the last decade and indicates further activities in conservation and the sustainable use of biodiversity in the coming 10–15 years. Long-term objectives of the national biodiversity conservation strategy are:

- Enhance the conservation function of protected areas including nature reserves, and establish protected area networks in order to achieve a complete national protected area system;
- Promote increased life quality for current and future generations by establishing a system of sustainable use of biodiversity.

The updated NBSAP has put forward strategic targets which promote sustainable economic development, prioritize conservation, and improve the eco-environment. The following actions were deemed a priority:

1. Establishment of a national protected area system that has effective management.
2. Implementation of a national protected area network.
3. Increased biodiversity conservation and management of natural parks in Mt. Kumgang and Mt. Chilbo.
4. Preparation of a Wetland Action Plan and restoration of degraded wetland ecosystems.
5. Update the "Red Data Book" and increase focus on conservation of threatened species.
6. Conservation of Black-faced spoonbill and cranes.
7. Community-based conservation and management of wildlife.
8. Develop ex-situ conservation of genetic resources.
9. Develop the National Biosafety Management Centre (NBMC).
10. Restore degraded forests and improve watershed management.
11. Showcase forest biodiversity conservation and its sustainable management.
12. Popularize agroforestry management.
13. Establish and disseminate environmentally friendly farming systems.
14. Combine organic farming and agricultural biodiversity conservation.
15. Increase coastal aquatic resources and promote sustainable use. Establish a monitoring system for coastal biodiversity.
16. Promote conservation and sustainable use of widely used traditional Koryo medicinal resources.
17. Establish eco-observation stations and observation networks to improve terrestrial ecosystem management.
18. Establish national biodiversity information systems.
19. Prepare biodiversity conservation plans at the provincial level.
20. Strengthen education, training, and public awareness of biodiversity conservation.

On the surface, the updated NBSAP looks like a brilliant master plan, which could be considered a great milestone for the DPRK in the implementation of its CBD and contribution to biodiversity conservation. However, very little information is available on accomplishments and progress, which suggests very little has been achieved in this updated NBSAP.

Reality–Two Cases

I have very little information on the implementation of the 10-year NFRP. 2017 is a benchmark year for the 10-year NFRP. By the end of 2017, the following goal should have been reached: planting trees in key sites such as highways, railways, villages, Taedong River Basin, and boundary areas. However, I did not see any new highway plantations, even though I had traveled around 150 km east and south of Pyongyang. In 2016, I unintentionally passed some plantations along a railway line, but not this time. Other than this I had not witnessed progress or accomplishments of any other targets.

In July 2017, I visited Songchon County, South Phyongan Province and Pongsan County, North Hwanghae Province. During these visits, I conducted interviews with officials and technicians at the county People's Committee and Director and County Forestry Management Board respectively. I visited nursery sites and planting sites. Table 8.3 provides background of these two counties.

Nursery. The County Forest Management Board allocated large staff resources to the nursery sites. In Songchon County, 300 out of 400 County Forest Management Board staff worked at nursery sites. There is roughly 70 ha of land designated for producing seedlings. Seeds were commonly used for producing seedlings of most species, including red

Table 8.3 Background information of Songchon County and Pongsan County

Name of the county		Songchon County	Pongsan County
Population		155,000	100,000
Rural population		49,000	35,000
Acrage of State forest land (ha)		46,000	25,000
Acrage of forest land to be afforested in 10-year NFRP		5,000 ha	3,000 ha
Acrage planted (ha)	2014		450
	2015		500
	2016	2,100	700
	2017	1,500	1020
	To be planted at 2018	700–800	700
Total number of staff in County Forest Management Board		400	300

pine, larch, acacia, chestnut, and maple. Cuttings were used for raising salix and popular seedlings. Sunshine covers are commonly used for protecting the younger seedlings of larch and maples. Seedlings were grown in very poor conditions. Taking popular seedlings as an example, at the end of July, the average seedling height was about 1.5 meters, and in China, they can grow to heights of 3–4 meters and produce seedlings for transplanting. Each county has at least 2 full-light spray nursery systems. These systems feature full-light and are used for the propagation of difficult-to-root seedlings, rather than normal seedlings. However, these full-light spray systems are used to propagate willows that are very easy to grow and propagate, and for the propagation of larch seeds. I introduced this all-optical spray system to Korea in 2006, and my employer has acquired all intellectual property rights. In 2006, two sets of spray system were introduced to the DPRK Central Nursery, and all counties now have at least two sets. It is important to adopt technology according to local conditions.

Plantation. I visited an afforestation site. In the low-lying hills, land was covered by various crops, including sweet potatoes, soybean, sesame, green bean, and corn, and also some younger trees such as paulownia, acacia, and larch, acer, which were grown in poor conditions. In the upper parts of the hills, there were some larch plantations with quite good survival rates. In Pongsan County, only 370 ha of planting area out of 2670 ha was left to be replanted and looked quite good in terms of plantation survival rate, as can be seen in Table 8.3. The director of the Songchon County forest management board said:

> In 2017, about 1,500 ha of barren land was afforested with 400 parcels of planting sites. The acrage of the largest and smallest planting site is 5 ha and 1 ha respectively. Everyone who is able should participate in planting including students, public servants, workers and peasants. Each institution, including schools, factories, hospitals and governmental agencies have been assigned a section of an afforestation site and is responsible for tree planting and seedling survival. Otherwise, institutions with financial capacity have to buy seedlings for planting at the request of the forest management board. Public institutions like schools and hospitals have to provide labour, and the County Forest Management Board provide free seedlings. In 2016 and 2017, the survival rate was high at 80% and 85% respectively, despite extreme draughts from April to early June 2017. The survival rate was good for acacia and chestnut seedlings, but relatively poor for red pine seedlings.—Field interview, conducted June 25th, 2017

Species selection for planting. Afforestation planning is designed by the MoLEP line agency, forest technology, and design institute. However, technicians consulted views from the local community. The fundamental principle for tree species selection is "right species to right site." However, the survival rate is another big concern as institutions involved in afforestation have a responsibility to ensure the survival of the planted tree seedlings. Species selection was not up to standard with regards to soil and water erosion control, biodiversity conservation, and other social-economic and environmental aims.

County Forest Management Board. This institute falls under the MoLEP system and is governed by the County People's Committee. At Pongshan County, there are 300 staff members, including 35 forest rangers, 30 forest policemen, and 20 administrative staff, with the rest working at nursery sites raising seedlings or involved in afforestation. The board supported peasants to plant fruit trees in their own gardens. In Songchan County, each peasant family had at least one jujube tree, which was in full bloom when I visited. Other roles of the board include:

(1) Allocating land for fuelwood to ri if required. According to regulations, each ri could be allocated 100 ha for planting trees for fuelwood. In Songchon County, 300 ha of forest land was designated land for animal grazing, and 4,000 ha of forest land was designated to ri to plant trees for fuelwood;
(2) Designing and planning forest restoration and management at a field and county level;
(3) Punish illegal collection and logging;
(4) Supervise the development of agroforestry (to be explained in detail in the next section).

During interviews, I was told the main difficulties faced by the board were:

(1) Insufficient seedlings. Large sites were required for raising seedlings;
(2) Insufficient materials to support the process of raising seedlings, such as a lack of pesticides, plastics, fertilizers, and transportation. Some sites had serious soil degradation, and so success with

planting was difficult to achieve, as it required biological and engineering measures. However key materials needed for engineering were lacking, such as cement and iron.

(3) Lack of transportation needed to deliver seedlings to planting sites. In Pongsan County, the director of the Board explained that the remaining afforestation sites were located far from the nursery, and had soil that was shallow and easily eroded. This meant there was no efficient way to deliver good seedlings to the site for planting.

None of the county board seemed to think controlling illegal harvesting was a pressing issue, even though it was causing severe forest degradation and deforestation. A director of the Board told me that every mountain entry point had been closed off and a large number of forest rangers and policemen were sent to patrol the forest area.

References

Agrawal, A., Chhatre, A., & Hardin, R. (2008). Changing governance of the world's forests. *Science, 320*, 1460–1462.

Antuma, J., Esch, B., Hall, B., Munn, E., Sturges, F. (2014). *Restoring forests and communities: Lessons from the Collaborative Forest Landscape Restoration Program*. School of Natural Resource and Environment, University of Michigan.

Aronson, J., & Alexander, S. (2013). Ecosystem restoration is now a global priority: Time to roll up our sleeves. *Restoration Ecology, 21*(3), 293–296.

Bastin, J. F., Finegold, Y., Garcia, C., Mollicone, D., Rezende, M., Routh, D., Zohner, C. M., & Crowther, T. W. (2019). The global tree restoration potential. *Science, 365*, 76–79.

Besseau, P., Graham, S., & Christophersen, T. (Eds.). (2018). *Restoring forests and landscapes: The key to a sustainable future*. Global Partnership on Forest and Landscape Restoration.

Chazdon, L. (2008). Beyond deforestation: Restoring forests and ecosystem services on degraded lands. *Science, 320*, 1458–1460.

Chazdon, L., Brancalion, P., Laestadius, L., Bennett, A., Buckingham, K., Kumar, C., Moll-Rocek, J., Guimarães Vieira, I., Wilson, S. (2016). When is a forest a forest? Forest concepts and definitions in the era of forest and landscape restoration. *Ambio, 45*(5), 1–13.

Clement, F. (2010). Analysing decentralised natural resource governance: Proposition for a "politicised" institutional analysis and development framework. *Policy Sciences, 43*(2), 129–156.

De Jong, W., Galloway, G. (2020). Narratives on the sustainable development goals and tropical forests. In J. Blaser & P. Hardcastle (Eds.), *Achieving sustainable management of tropical forests*, in press. Burleigh Dodds Science Publishing.

Erbaugh, J., & Oldekop, J. (2018). Forest landscape restoration for livelihoods and well-being. *Current Opinion in Environmental Sustainability, 32*, 76–83.

FAO. (2010). *Global forest resource assessment report.*

Gilmour, D. (2005). An historical account of fuelwood restoration efforts. In S. Mansourian, D. Vallauri, & N. Dudley (Eds.), *Forest restoration in landscapes: Beyond planting trees* (pp. 223–227). Springer.

Görg, C. (2007). Landscape governance: The "politics of scale" and the "natural" conditions of places. *Geoforum, 38*(5), 954–966.

Guariguata, M., & Brancalion, P. (2014). Current challenges and perspectives for governing forest restoration. *Forests, 5*(12), 3022–3030.

Jin, Z., Huizhi, Z., & Mingxu, C. (2016). Environmental management and sustainable development in Korea. *Population Journal, 4*, 88–92 (in Chinese).

Hobbs, J. R., Higgs, E., Hall, M. C., Bridgewater, P., Chapin, S. F., III., Ellis, C. E., Ewel, J. J., Ewel, J. J., Hallett, L. M., Harris, J., Hulvey, K. B., Jackson, S. T., Kennedy, P. L., Kueffer, C., Lach, L., Lantz, T. C., Lugo, A. E., Mascaro, J., Murphy, S. D., Nelson, C. R., … Yung, L. (2014). Managing the whole landscape: Historical, hybrid, and novel ecosystems. *Frontiers Ecology Environment, 12*, 557–564.

Katila, P., Colfer, C. J. P., de Jong, W., Galloway, G., Pacheco, P., Winkel, G. (Eds.). (2019). *Sustainable development goals: Their impacts on forests and people*. Cambridge University Press.

Lamb, D., Erskine, P. D., & Parrotta, J. A. (2005). Restoration of degraded tropical forest landscapes. *Science, 310*(5754), 1628–1632.

Laurance, W. F., Sayer, J., & Cassman, K. G. (2014). Agricultural expansion and its impacts on tropical nature. *Trends in Ecology & Evolution, 29*, 107–116.

Li Shixian. (1985). Management of Forest Industry in Korea. *Forestry Economy, 1*, 61–63 (in Chinese).

Long, H., Liu, J., Tu, C., & Fu, Y. (2018). From state-controlled to polycentric governance in forest landscape restoration: The case of the ecological forest purchase program in Yong'an municipality of China. *Environmental Management, 62*(1), 58–69.

Mansourian, S., & Vallauri, D. (2014). Restoring forest landscapes: Important lessons learnt. *Environmental Management, 53*(2), 241–251.

Mansourian, S., Vallauri, D., & Dudley, N. (2005). *Forest restoration in landscapes: Beyond planting trees*. Springer.

MoLEP. (2011). *DPRK 4th report on CBD.*

Monroe, A. S., & Butler, W. H. (2016). Responding to a policy mandate to collaborate: Structuring collaboration in the collaborative forest landscape restoration program. *Journal of Environmental Planning and Management, 59*(6), 1–19.

Newton, A. C., Del Castillo, R. F., Echeverría, C., Geneletti, D., González-Espinosa, M., Malizia, L. R., Premoli, A. C., Benayas, J. M. R., Smith-Ramirez, C., & Williams-Linera, G. (2012). Forest landscape restoration in the drylands of Latin America. *Ecology and Society, 17*(1). https://doi.org/10.5751/ES-04572-170121

Ostrom, E. (2010). Polycentric system for coping with collective action and global environmental change. *Global Environmental Change, 20*, 550–557.

Ros-Tonen, M. A. F., Derkyi, M., & Insaidoo, T. F. G. (2014). From co-management to landscape governance: Whither Ghana's modified taungya system? *Forests, 5*(12), 2996–3021.

Sabogal, C., Besacier, C., & McGuire, D. (2015) Forest and landscape restoration: Concepts, approaches and challenges for implementation. *Unasylva, 245, 66*(3), 3–10.

Sayer, J., Chokkalingam, U., & Poulsen, J. (2004). The restoration of forest biodiversity and ecological values. *Forest Ecology Management, 201*, 3–11.

Sayer, J., Sunderland, T., Ghazoul, J., Pfund, J. L., Sheil, D., Meijaard, E., ... & Buck, L. E. (2013). Ten principles for a landscape approach to reconciling agriculture, conservation, and other competing land uses. *Proceedings of the National Academy of Sciences of the United States of America, 110*, 8349–8356.

Seymour, F., & Busch, J. (2017). *Why forests? Why now? The science, economics and politics of tropical forests and climate change*. Center for Global Development.

SOFO. (2018). *State of the World's forest 2018*. Food and Agricultural Organiztion.

Van Oosten, C., Gunarso, P., Koesoetjahjo, I., & Wiersum, F. (2014). Governing forest landscape restoration: Cases from Indonesia. *Forests, 5*, 1143–1162.

Van Oosten, C., Runhaar, H., & Arts, B. (2021). Capable to govern landscape restoration? Exploring landscape governance capabilities, based on literature and stakeholder perceptions. *Land Use Policy*. https://doi.org/10.1016/j.landusepol.2019.05.039

West, P. C., Gerber, J. S., Engstrom, P. M., Mueller, N. D., Brauman, K. A., Carlson, K. M., Cassidy, E. S., Johnston, M., MacDonald, G. K., Ray, D. K., & Siebert, S. (2014). Leverage points for improving global food security and the environment. *Science, 345*, 325–328.

Xu, J., Noordivijk, M., He, J., Kim, K.-J., Jo, R.-S., Pak, K.-G., Kye, U.-H., Kim, J.-S., Kim, K.-M., Sim, Y.-N., Pak, J., Song, K.-U., Jong, Y.-S., Kim, K., Pang, C.-J., & Ho, M. (2012). Participatory agroforestry development for restoring degraded sloping land in DPRK. *Agroforestry Systems, 85*(2), 291–303.

CHAPTER 9

Agroforestry Development in the DPRK

It was widely accepted that agroforestry is an important means of achieving sustainable development and resolving the ecological and environmental problems in rural areas. The DPRK is highly populated with many mountainous and hilly areas in its territory. Arable land is limited and has been in decline due to destruction caused by natural disasters and construction. Lack of fertilizers has also resulted in food shortages. The DPRK only had one option—to produce more food in forest land—and that is the fundamental reason why agroforestry was incorporated into the national strategy and seen as a priority for economic development.

Forests Were Traditionally Part of Korean Life and Production

Northeast Asia has 5,000 years of recorded history and civilization, during which time its people had lived in close proximity to forests, practicing hunting; gathering wild food, fuelwood, medicinal plants, and other non-timber forest products; and engaging in forest-based agriculture. People of this region have acknowledged the value of forest resources, as evidenced by their rich traditional forest management practices. The most common of these practices at the community level includes managing common forests or village groves: fengshui forests (in China), maeulsoop (in Korea), and satoyama (in Japan).

L. Jinlong, *Crying Forests*,
https://doi.org/10.1007/978-981-16-9725-8_9

Maeulsoop Woods in Korean Villages

The Republic of Korea constructed village woods called maeulsoop or bibosoop in order to live in harmony with the land (Kim & Jang, 1994; Lee, 2008). This culture is somewhat related to the philosophy of feng-shui, a practice which claims that a place surrounded by high and firm mountains to the north and relatively gentle but uninterrupted ranges to the west and eastern sides is a good place to live, and so people feel most comfortable in the middle. The end of a range forms an outlet of village watershed. The outlets must be lower than the mountain, but should also have some height and be interconnected with each other to shield the village. In reality, however, these conditions often do not exist. If people thought there were some defects in the environment of their village, they complemented the defects by constructing village woods. The most common practice was to establish woodlands in the outlet of the village watershed. When the outlet was opened, they planted trees in the form of a grove to protect the village and to make it a good place to live. The canopy of groves functions like a curtain. From the inner side of the village woods, people can look at the outer world while from the outer side, a passerby hardly sees the village. It is also believed that the grove keeps the evil spirits from entering the outer world. People protect these village groves and traditionally use them for village meetings to discuss community affairs.

Village woods have various functions: for religious purposes as shrine woods; as windbreaks or shelter belts; for erosion control; and for esthetic beauty. Take a typical well-maintained village wood in Dundong village, Hwasoon County, Cheonnam Province, South Korea as an example (Youn et al., 2011). In Dundong village, the people used manpower to construct a bank along the river in front of the village to protect it against the water flow, and then planted trees on the bank. The village woods became effective in protecting against strong winds and floods, and in keeping the village out of sight from strangers. The village woods also screen people's eyes from ugly looking rocks. In spiritual terms, the woods help in the upbringing of youngsters in a manner that makes them see good things and makes their minds beautiful.

Many maeulsoop woods in the DPRK have been lost or degraded over the last century (Youn, 2009). The underlying causes of village grove loss in both North and South Korea include total change of land ownership, erosion of traditional culture and religion in the name of modernization.

The privatization of forestland initiated by the First Forest Law in 1908 is considered to be one of the most important institutional changes that contributed to the loss of village woods in Korea. There has been no such cultural tradition to support maeulsoops, which also depend on the existence of cultural capital such as traditional rituals based on natural worship and community organization in charge of woodland management.

Traditional Management of Forests

The use of charcoal has a long history in Northeast Asia. A document of the ancient Shilla Korean kingdom which dates back to the seventh century, records the use of charcoal for heating the houses in Gyeongju, the kingdom's capital.

Prior to colonization by Japan in 1910, the forest policy of kingdoms of the Korean Peninsula maintained public ownership of forest and wetlands. Even though the entire forestland of the Korean Peninsula was proclaimed belonging to the king, some parts of forestland were particularly distributed to royals and high officials. As commerce and the market economy developed in the seventeenth and eighteenth centuries, many of the forestlands were occupied by noble families and rich merchants through the establishment of private graveyards in the forests. As the private occupation became more apparent, less forestland was left from which poor villagers could collect fuelwood and timber for housing. The villagers formed a club, with representatives from every village household to manage a common forest, not privately occupied by powerful families. The community-based forest management club was called Song-gye, meaning "club of pine forest," which signifies the dominant tree species of the forestland at the time. Song-gye, which was based on a by-law of the community for managing the common forest, organized various activities such as controlling wildfires and illegal logging. It has played an important role in the formulation and implementation of forest policy for the rehabilitation of denuded forest landscapes in recent decades in South Korea (Chun & Tak, 2009).

Traditional Cultural and Religious Practices Related to Forests

The psychological or spiritual effects of forests on human beings are reflected first in people's religious beliefs (Parrotta et al., 2009). Trees and forests are the foundation of various folk religions. By worshiping trees and forests, people attempted to draw comfort for their mental and physical weaknesses. "Seo Nang Dang," a Korean village shrine, is a place for the God protecting the village. It lies in the forest at the entrance to the village. It consists of a "Dangzip," where the God resides, and a holy tree called "Seonang Namu," the passageway for the god to descend from heaven. The God governs the villagers' fortune and prosperity, and for that reason each village holds a religious ritual at the start of the year, which is also a festival day for the village. Through these rituals, Seo Nang Dang worked to tie the villagers together spiritually. Such folk religions have faded away as they clashed with the modern world, and new religions spread out and replaced them. Nonetheless, Seo Nang Dang is a definitive trace of a folk religion related to trees and forests. It represents forest cultures of the past, which played a critical role in determining the composition of the nature-cultural landscape of the countryside.

In the Korean Peninsula, "Seongjushin" is the house guardian God who protects the house, and it is believed that he resides in a pine tree (Pinus *densiflora*). There is a legend saying that this God used to live in heaven, but after committing wrongful deeds he was sent to the world of humans. There, he found out that he had nowhere to live, so he prayed to the Heaven God, and the Heaven God allowed him to get seeds of a pine tree from a temple. He planted these seeds and then cut down the grown-up pine trees to build his house. That is why Seongjushin is believed to be living in pine trees. The people there perform a small ritual for Seongjushin by fixing in girders while building their houses. This legend signifies the importance of growing trees for the people living in the Korean Peninsula to be able to live in a comfortable house made of timber.

Initiation of Agroforestry

Agroforestry Development in Slope Land to Alleviate Hunger

The DPRK is a mountainous country. Under the influence of Eastern European socialist countries, it has committed to agricultural modernization. The DPRK had achievements related to agricultural modernization before the collapse of Eastern Europe. In the past, the DPRK learned from the Great Stalin Plan for Reconstruction of Nature,[1] implementing large-scale afforestation projects across wide-scale forest belts in windy land degradation areas on the plateau. The projects have improved the agricultural production conditions in the plateau. Ginseng, various herbal medicines and mushrooms like matsutake, have also been widely collected. After the 1950s, the DPRK restored forest vegetation on the identified forest land, regardless of the function of food production.

However, Eastern Europe collapsed in 1991. Since Russia stopped the annual supply of 100 thousand tons of oil, the DPRK faced a new challenge with energy. Also, severe natural disasters and food crises have caused the method of slash-and-burn, that had been stopped for more than two decades, to resurge and expand again. However, the open-field planting was not counted in farm land, and planting requires the payment of grain quotas (10% to primary and secondary schools, 40% of the output must be paid to the country, and the rest distributed among members of the commune).

The DPRK suffered from food and energy shortages and large-scale deforestation in the 1990s (Tang et al., 2010). As a consequence, agroforestry has become a management practice in the sloping lands of the DPRK. It is a bottom-up participatory process of exploring new opportunities for the development of local technology and foods for survival. As stated by Xu and his colleagues (Xu et al., 2012), all three rights—right-to-use, right-to-harvest, and right-to-plan—were new in the DPRK and jointly contributed to success, as there was increased access to sloping lands for tree products and food, which were used by groups in the pilot sites. They also highlight the growth of technological innovation once social and institutional constraints to land access were removed. However, it was amazing to me that agroforestry prevailed in the DPRK, where top-down planning and a so-called "scientific" approach to industrialized agriculture dominate (Park, 2009).

[1] In order to ensure high and stable production in some grasslands and forest-steppe areas of the Soviet Union in Europe, the former Soviet Union planted 5.7 million hectares in the desert areas of Central Asia from the north to the south, and constructed 8 shelterbelts running through the Russian prairie to form a huge windbreak forest belt. In order to alleviate the hazards of wind and sand. This is the world's largest ecological environment project, which corresponds to the Great Plains Shelterbelt in North America (the Roosevelt Project), which the former Soviet Union used to call the Great Stalin Plan for Reconstruction of Nature.

Members of the county forest management committee actually come from various native places, and they also faced severe food shortages. Employees of county forest management committees often turned a blind eye and pretended not to see the mass deforestation and large-scale reclamation of wasteland by locals for food production purposes. This sort of illegal farming was not reported in official documents, and naturally there were no requirements to share food products to public agencies, including the government. As early as 2013, MoLEP planned to formulate a state law to regulate these activities, legalize the land that was being illegally harvested, and asked villagers using the land to turn in 10% of the grain harvest.

I strongly recommended the MoLEP officials reconsider this proposal and that no quota was collected from this "illegal" farming. In fact, villagers that had reclaimed forest land for private farming did not use any fertilizers. Harvesting of crops was down to luck and yield was extremely low. In a good weather year, average grain yield was only 3–4 tons per hectare. If severe droughts and floods occurred, there was often no harvest.

I was worried at the fact that as long as the law permitted local officials to collect taxes and fees from locals, and this provided an incentive for grassroot officials to intervene in agroforestry production. This would definitely hurt the motivation of agroforestry producers, condone micro-corruption, solidify the interest class, and will further exacerbate the famine in the DPRK.

Adopted by Government Authorities

In 2000, the population of cities such as Pyongyang remained unchanged, but the urbanization rate had declined. Some residents in small and medium urban areas were transferred to mountain areas. That is to say, more mountainous forests were opened up as permanent farmland. I presume these people suffered, as they had to adapt a new way of life and had production quotas, which brought about many hardships. According to my personal estimates, of 9 million hectares of forest land, only about 5 million hectares of forest remained. Among the 4 million hectares of forest land, at least 1 million hectares had been illegally reclaimed for planting food.

In order to reverse degradation, increase yields and generate more income from mountains and hills; the SLM project was introduced by the Ministry of Land and Environmental Protection (MoLEP) in the early 2000s. With the support of the Swiss Development Agency in 2000 and the United Nations Food and Agriculture Organization in 2004, agroforestry was increasingly accepted by the DPRK technical department. Meanwhile, slope farmland (forest land) was successfully established and achieved good performance. The 1st and 2nd National Workshops on Agroforestry Management held in July 2008 and October 2009 provided good opportunities to exchange information and share ideas on the development and introduction of agroforestry technologies. Table 9.1 provides more information on a sloping-land management project at Suan County, North Hwanghae Province, where national policy reforms were being discussed. By 2011, there were more than 87 sloping land user groups in North Hwanghae Province, and several hundred other user groups elsewhere in the country. Tree cover on these lands had increased, and land productivity was substantially higher than it was when used by individual squatters.

After 2010, driven by the Ministry of Land and Environment Protection, agroforestry was accepted by DPRK leaders and became an important national strategy. With the permission of the upper-level government, laws, technologies, and management systems were quickly established.

Agroforestry Strategy and Action Plan in the DPRK

With the financial support of the Swiss Agency for Development and Cooperation and Technical support from ICRAF (World Agroforestry Centre), the DPRK developed a national agroforestry strategy and action plan in 2013.

The DPRK government has defined agroforestry as an important economic strategy to provide food and reforest the mountains and to contribute to overall environmental protection and sustainable economic development in the long run. The general goal of the agroforestry strategy is to contribute to the construction of a prosperous socialist state by improving food security and people's livelihoods, while restoring the country's landscape and environment through the active implementation of agroforestry in all provinces, cities and counties, which has 6 specific goals:

Table 9.1 Timeline of events for the sloping-land management project at Suan County in northern Hwanghae Province

Time	*Natural hazards and risks*	*State responses*	*International supported Project activities*
1996	Floods	State called for land management movement	
1997	200–300 forest fire incidents caused by slash-and-burning		
1999		State policy on afforestation and greening in September and followed by the 10-year plan provincial initiative for agroforestry development to contribute to national forest policy	
2000	Pine caterpillar outburst, extreme cold/freezing in winter, damage in persimmon		
2002	Serious Flood		
2003		Establishment of SLUGs	SDC-MOLEP inception workshop on SLM Identifying pilot sites
2004		3 user groups established in Suan	
2005	Extreme drought caused loss of 60% of seedlings in the nursery and field expansion of project sites into seven user Groups		

Time	*Natural hazards and risks*	*State responses*	*International supported Project activities*
2006		12 user groups	
2007			Project implementation handed over to MoLEP Humanitarian aid used for development cooperation
2008		Establishment of pest control groups Additional six user groups in Hwangju County	1st national agroforestry workshop
2009			Expansion into 34 user groups

(*Source* Xu et al., 2012. revised. And project funded by the Swiss Agency for Development and Cooperation (SDC))

- Adopt agroforestry in non-forested and low productive forest lands, to achieve food security and create a reliable base for the production of various raw materials while accelerating forest ecosystem restoration;
- Implement agroforestry on sloping agricultural lands, improve forest cover, and create a sustainable and productive environment while maintaining food security;
- Reduce disaster risk as much as possible, protect overall environment, ensure a healthy environment for agriculture production, and reduce dependence on forests, by planting lots of trees in disaster-probe areas such as sloping arable lands and riparian lands.
- Raise public agroforestry awareness and capacity, making agroforestry extension the work of everyone and the entire society.
- Constantly improve economic profitability and the effectiveness of agroforestry in environmental protection, and strengthen the scientific basis by doing more research into agroforestry.
- Strengthen institutional capacity and provide a legal and administrative basis for the development of agroforestry, by strengthening coordination and collaboration between relevant stakeholders and improving relevant laws and regulations.

In the action plan, there were clear meaningful goals for action: (1) intercrop trees and crops on 300,000 ha of forest land cleared for farming, to contribute to food production and reforestation.

Agroforestry Technology

Agroforestry Models in the DPRK

There are many kinds of agroforestry practices now in the DPRK which can be categorized as follows:

- **Windbreak shelter-belt forest**. Influenced by the former Soviet Union in 1952, windbreak shelter-belt forests were established in the Northern highlands, which assist in claiming thousands of hectares of abandoned land for growing potatoes. That was extended to plateaus area, which improved farming conditions greatly. In 2004, with the assistance of the FAO, some windbreak shelter-belt forests were established in paddy field areas, which aimed to provide fuelwood to rural households and reduce fuelwood harvesting in the Neighbor Mountains.

- **Silvi-pastural systems**. Some forests were opened for grazing, so that forests and livestock could co-exist in degraded areas.
- **Home Gardens**. Rural households had a small piece of land which could be used for independent production and self-subsistence. There was great potential to produce fruit or cash trees in local gardens.
- **NTFPs under forest canopy**. Medicinal herbs and mushrooms could be cultivated under the forest canopy of Platycodon grandiflorum, Codonop sispilosula, and Lentinus edodes.
- **Inter-cropping**. Due to severe food shortages, slash-and-burn farming made a reappearance again, which caused large-scale destruction of forests. From early 2000, the number of international organizations, including the Swiss Development Cooperation (SDC) and FAO, has demonstrated inter-cropping in sloping lands to achieve twin goals of increasing food production and preventing soil and water erosion.

Tree Species Selection for Sloping-Land Agroforestry

In the DPRK, local people paid no attention to tree species selection. They were not involved in the selection process at all as it was dominated by professionals from state forest agencies, who did not pay attention to local needs. Species were usually selected according to a single criterion—economic or environmental benefit, focusing on fast-growing timber species including larch and pine (Xu et al., 2012). However, the top-down approach for agroforestry tree species selection has proven ineffective (Franzel et al., 1995; German et al., 2006; Reubens et al., 2011). In agroforestry for sloping-land management, the choice of species is highly complex in both biological and socioeconomic terms (Kumtashula & Mafongoya, 2005). From an ecological perspective, the erosion and runoff control provided by a mix of appropriately selected and placed trees can help achieve watershed and biodiversity conservation objectives. For socioeconomic reasons, successful selection of tree species in mountainous areas can increase agricultural productivity and supplement farmer income through the sale of fruits, nuts, fuelwood, and timber, while mitigating fuelwood and timber shortfalls. There was no simple and practical way to identify the social and environmental trade-offs in tree species selection (German et al., 2006; He et al., 2015).

Participatory methods hold the greatest potential to involve farmers in the design of agroforestry systems (Haggar et al., 2001; Hu et al., 2015). Tree species for sloping-land agroforestry was chosen to address the perceived needs of locals. Priority setting and characterization based on user perception is an essential part of tree species selection. User groups did not actively take part in land use management if plans didn't appear to improve their livelihoods. There was a particular need for tree selection to serve both environmental and food security purposes in the DPRK.

As farmers and communities know their needs and local site conditions best, participatory approaches that involve farmers can help to generate economically and environmentally sound technologies and increase sustainability of natural resource management. Hu and his colleagues (2015) demonstrated that participatory approaches enabled local communities to select multipurpose species, which differed from government species selection that focused on timber and fast-growing species. As a result of participation, this study was able to select multipurpose tree species for agroforestry in sloping-land management in 3 counties (more details in Table 9.2).

During a 2012 review workshop, government officials expressed their appreciation of the participatory approach of this SDC pilot project (Xu et al., 2012), as it advanced agroforestry development in 5 practical respects: (1) Responded to problems, needs, and opportunities identified by users; (2) Identified and evaluated technology options that build on local knowledge and resources; (3) Ensured technical innovations were appropriate for local socioeconomic, cultural, and political contexts; (4) Promoted wider sharing and use of agroforestry innovations; (5) Built capacity and promoted knowledge exchange.

In 2013, the DPRK government launched the National Agroforestry Policy, and in 2014, it formulated a participatory approach as a key strategy for forest ecosystem restoration and local livelihood development. Policy requirements meant local participation in agroforestry species selection was necessary in order to achieve both economic and ecological benefits in sloping-land management. However, without external support, participatory agroforestry in other parts of the DPRK were all talk and no action.

Table 9.2 County-specific choices of species for agroforestry in SIDA sites

Main benefit	*Suitable location*	*Preferred species*	*Counties*
Fruit	Agroforestry near villages	Prunus armeniaca, Castanea crenata, Ziziphus jujuba, Pyrus ussuriensis	Suan
		Hippophae rhamnoides Sohung	Yonsan
Erosion control, fuelwood	Forest and agroforestry on sloping land	Alnus japonica, Salix gracilistyla, Juglans mandshurica	Suan
		Ulmus pumila	Suan, stony area
		Morus alba	Suan, Sohung
Timber Forest	On mountain top	Larix olgensis var. koreana, Larix leptolepis, Kalopanax pictum, Tilia amurensis	Suan
Other income	Agroforestry on sloping land	Pinus koraiensis, Evodia daniellii, Xanthoceras sorbifolia	Sohung
		Quercus acutissima, Aronia melanocarpa	Suan, Yonsan

(*Source* He et al, 2015; Hu & Xu, 2017)

Other Agroforestry Issues in the DPRK

Institutional Establishment for Promoting Agroforestry

In March 2013, a central non-standing agroforestry committee was composed of officials from MoLEP, MOA, MoTI, the government agency of education, science, and technology, the State Planning Committee, the Ministry of Food and Consumer Stuff Industry and the Ministry of Chemical Industry. The committee operated under guidance of the Cabinet and had a role in planning, coordinating, and assessing annual agroforestry development across the country, and also proposed measures to the Cabinet, such as changing legislative framework in order to accelerate agroforestry development.

Not long after, non-standing agroforestry committees were organized at every provincial, city, and county level authority. Agroforestry divisions have been set up in MoLEP, MoA, and MoFI. A nationwide administration system for agroforestry has also been established.

Science and education bodies related to agroforestry were established in 2013. Agroforestry research and design units were set up in the Academy of Forest Sciences under MoLEP. The agroforestry research unit was established in the Academy of Agricultural Science under MoA. From 2013 onwards, an agroforestry curriculum was developed. The curriculum was compulsory in forestry, agriculture, and land management faculties in universities specializing in agriculture and forestry and provided professional training on agroforestry to undergraduates and technicians. In July 2014, the Korean Agroforestry Technical Association was established under the Korean General Federation of Science and Technology, which facilitated agroforestry experimentation, research, and extension.

Regulation

In May 12, 2013, Cabinet Resolution No. 35 on Agroforestry Regulation was published. In article 12, "forest land use licenses" can be issued to user groups and individual households, and the use of illegally cleared forest sloping lands for agroforestry practices was legalized. Motivations and interests of user groups and individual households was ensured and the government committed to providing material and technical support to maximize output from sloping lands.

Challenges Faced

Although local people were allowed to use 300,000 hectares of slopes for 3–7 years, forests in the DPRK are actually state-owned, similar to Russia, and so once trees were planted on slopes by government departments, locals had restricted land use. After trees were fully grown, they too were owned by the government but farmers could harvest food grown in forests. However, the survival rate was low, generally, only 20–30% and so the likelihood of trees actually growing into forest was also very low. Government agencies continually interfered with land planted by farmers, which was a huge challenge.

Similarly, problems emerged planting trees on agricultural land, as the local community had no incentive to plant trees, especially on gentle slope agricultural land.

This was because sections of land had to be occupied, which affected food production.

The relationship between the interests of local communities and forest management had not been established.

Each county had a set area of agricultural forest. In Pongsan County, a total of 600 hectares was approved for agroforestry development, owned by about 1,000 members, and managed by land user groups. The team using the land was responsible for afforestation, and the seedlings were purchased by members at the county nursery. Fast-growing tree species could only be inter-cropped for 5–7 years, while slower-growing tree species could be inter-cropped for 8–10 years. This model was an example of early inter-cropping. MoLEP actually reduced expenditure and achieved afforestation goals, which was better than the mobilization afforestation model. However, it was not the agricultural forestry development that we had envisioned.

References

Chun, Y. W., & Tak, K.-I. (2009). Songye, a traditional knowledge system for sustainable forest management in Choson Dynasty of Korea. *Forest Ecology and Management, 257*, 2022–2026.

Franzel, S., Hitimana, L., & Akyeampong, E. (1995). Farmer participation in on-station tree species selection for agroforestry: A case study from Burundi. *Experimental Agriculture, 31*(1), 27–38.

German, L. A., Kidane, B., & Shemdoe, R. (2006). Social and environmental trade-offs in tree species selection: A methodology for identifying niche incompatibilities in agroforestry. *Environment, Development and Sustainability, 8*(4), 535–552.

Haggar, J., Ayala, A., & Dı´az, B., & Reyes, C. U. (2001). Participatory design of agroforestry systems: Developing farmer participatory research methods in Mexico. *Development in Practice, 11*(4), 417–424.

He, J., Ho, H., & Xu, J. (2015). Participatory tree species selection for sloping land management in North Korea. *Mountain Research and Development, 35*(4), 318–327.

Hu, J., & Xu, J. (2017). Is there decentralization in North Korea? Evidence and lessons from the Sloping land management program 2004–2014. *Land Use Policy, 61*, 113–125.

Kim, H. B., & Jang, D.-S. (1994). *Maulsoop, the Korean village grove* (p. 204). Yeolhwadang Publishing Co (in Korean).

Kumtashula, E., & Mafongoya, P. L. (2005). Farmer participatory evaluation of agroforestry trees in eastern Zambia. *Agricultural Systems, 84*, 39–53.

Lee, D. (2008, October 5–10). A Korean perspective of traditional knowledge forest-related knowledge. In *Proceedings of the 1st international conference on forest related traditional knowledge and culture in Asia* (pp. 3–5). Korea Forest Research Institute.

Park, P. H. (Ed.). (2009). *The dynamics of change in the DPRK: An institutionalist perspective*. Institute for Far Eastern Studies, Kyungnam University.

Parrotta, J. A., Lim, H. F., Liu, J., Ramakrishnan, P. S., & Youn, Y.-C. (2009). Traditional forest-related knowledge and sustainable forest management in Asia. *Forest Ecology and Management, 257*, 1987–1988.

Reubens, B., Moeremans, C., Poesen, J., Nyssen, J., Tewoldeberhan, S., Franzel, S., Deckers, J., Orwa, C., & Muys, B. (2011). Tree species selection for land rehabilitation in Ethiopia: From fragmented knowledge to an integrated multi-criteria decision approach. *Agroforestry Systems, 82*(3), 303–330.

Tang, L., Shao, G., Piao, Z., Dai, L., Jenkins, M. A., Wang, S., Wu, G., Wu, J., & Zhao, J. (2010). Forest degradation deepens around and within protected areas in east Asia. *Biology Conservation, 143*(5), 1295–1298.

Xu, J., Noordivijk, M., He, J., Kim, K.-J., Jo, R.-S., Pak, K.-G., Kye, U.-H., Kim, J.-S., Kim, K.-M., Sim, Y.-N., Pak, J., Song, K.-U., Jong, Y.-S., Kim, K., Pang, C.-J., & Ho, M. (2012). Participatory agroforestry development for restoring degraded sloping land in DPRK. *Agroforestry Systems, 85*(2), 291–303.

Youn, Y.-C. (2009). Use of forest resources, traditional forest-related knowledge and livelihood of forest dependent communities: Cases in South Korea. *Forest Ecology and Management, 257*, 2027–2034.

Youn, Y.-C., Liu, J., Daisuke, S., Kim, K., Ichikawa, M., Shin, J.-H., & Yuan, J. (2011). Northeast Asia (Chapter 8). In J. A. Parrotta & R. L. Trosper. (Eds.), *Traditional forest-related knowledge: Sustaining communities, ecosystems and biocultural diversity* (pp. 281–313, Vol. 12). World Forest Series, Springer.

CHAPTER 10

From Afforestation to Forest Landscape Restoration in the DPRK: Gaps and Challenges

The DPRK is currently in a state of severe deforestation and forest degradation, such that it was selected as a beneficiary for Forest and Landscape Restoration (FLR) programs run by the Food and Agriculture Organization of the United Nations (FAO). This study investigated the status and gaps in the implementation of FLR action and analyzed the key challenges to FLR implementation. To accomplish this, we visited ministries relevant to forest and landscape restoration, selected and interviewed officials and professionals, and based our research on 10 years of experiences working in the DPRK. Further, we propose a roadmap for FLR implementation. Our results indicate that there is a need for a profound shift in attitudes toward the role of local people, recognizing that local people can be the main actors and partners in forest and landscape management. Policy reform is needed in the forestry sector to support the transformation of local livelihoods into a forest-friendly paradigm and to exit the vicious cycle of poverty and hunger related to environmental deforestation and degradation. Future FLR policies need to focus on the decentralization of forest management, clarification of forest tenure, and improvement of multi-sectoral cooperation and multi-level collaboration.

L. Jinlong, *Crying Forests*,
https://doi.org/10.1007/978-981-16-9725-8_10

Forest Landscape Restoration for Breaking the Vicious Cycle

Deforestation and forest degradation are closely aligned with issues of poverty, food insecurity, climate change, and biodiversity loss, and they continue to persist as a primary political, social, and environmental challenge in the DPRK (Laurance et al., 2014; West et al., 2014). According to various sources, the rate at which forests are cleared or degraded in the DPRK is unclear. However, according to projections published in the Global Forest Resource Assessment by the FAO (FAO, 2010), the data indicates a rapid decline in forest cover, from 8.20 million hectares in 1990 to 5.67 million hectares in 2010, representing a deforestation rate of more than 2% per year, and a decline in growing stock from 526 to 330 million m^3, at an approximate rate of 3% per year over the past two decades. The DPRK is facing serious deforestation and forest degradation. According to my best estimates, approximately 1.5 million ha of forests on sloping land have changed to farming or shifting cultivation since 1993, with the complication that most of these areas have a high slope, consequently resulting in widespread erosion and landslides owing to heavy rainfall. To some extent, the DPRK demonstrates a vicious cycle of deforestation and forest degradation, with interconnected issues of poverty, food and fuelwood shortage, loss of biodiversity, and poorer resilience as shown in Fig. 10.1. The DPRK is an important region for

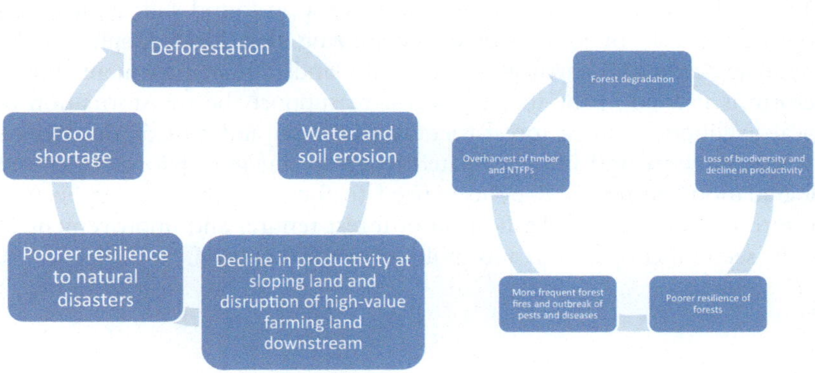

Fig. 10.1 Vicious cycle of deforestation and forest degradation in the DPRK

the conservation of biodiversity owing to its relatively abundant biodiversity and geographical position that connects the continent and ocean. Threatened plant species (seed plant) account for approximately 4.54% of the total number of species. Threatened animal species comprise approximately 21% of all vertebrate species (Ministry of Land and Environment Protection, 2011).

There is growing acknowledgment that conventional sectorial approaches of afforestation to address these often interconnected challenges are insufficient (Tscharntke et al., 2012), such that holistic and integrated approaches—forest landscape restoration (FLR)—are recommended (Reed et al., 2015; Sayer et al., 2014; Van Oosten, 2013). FLR is defined as "a planned process that aims to regain ecological integrity and enhance human well-being in deforested or degraded forest landscapes" (IISD, 2002), which emphasizes the importance of approaching land management as something that is broader and more comprehensive than a set of site-level technical interventions (Sabogal et al., 2015). It has also been emphasized as an action for integrated and holistic approaches for intervention in complicated issues, in particular for opportunities that require the balancing of trade-offs or those that need to deliver multiple benefits simultaneously at the landscape level in the context of climate change and social and economic transformation (Bollenbacher et al., 2014). FLR seeks a balance between restoring ecosystem services related to wildlife habitats, biodiversity, water regulation, and carbon storage and supporting the productive function of land for agriculture and other related uses by considering landscapes as a mosaic of ecosystems and forests as part of the livelihood system of inhabitants. The more widely accepted definition proposed by the Global Partnership on Forest and Landscape Restoration (GPFLR) is that FLR is "an active process that brings people together to identify, negotiate, and implement practices that restore an agreed optimal balance of the ecological, social, and economic benefits of forests and trees within a broader pattern of land uses" (GPFLR, n.d.). This definition emphasizes that FLR is a concept that refers to an approach that involves stakeholders in all affected land use sectors and applies participatory decision-making processes (Sabogal et al., 2015). Although there is no universally accepted concept on FLR and disagreement over terminology in a large number of studies, the application of forest and landscape restoration is part of the international agenda, and state forest policies in many countries (Sayer et al., 2013; Scherr et al., 2013). At a GPFLR conference in Bonn, international communities, including the International Union for Conservation of Nature, the Convention on Biological Diversity (CBD), and the UN Framework Convention on Climate Change (UNCCC), committed to a restoration target of 150 million ha (called the Bonn Challenge). Moreover, Asia-Pacific Economic Cooperation leaders pledged to increase their forest area by 20 million ha by 2020 (FAO/RECOFTC, 2016: 3). In 2014, the FAO launched the "Forest and Landscape Restoration Mechanism" (FLRM) to support countries in the planning and implementation of FLR as a means to achieve the Bonn Challenge. The DPRK was selected as a pilot state among the nine countries supported by the FLRM, aiming to restore 1.5 million ha of lost forests and degraded lands.

In the DPRK, where forests comprise approximately 80% of the total land area, and water bodies comprise 6%, the management of mountains (shan) and water (shui) was highly prioritized for socioeconomic development, livelihood improvement, food security, and addressing environmental concerns (including climate change and biodiversity loss). As shown in Fig. 10.1, deforestation and forest degradation are interconnected with food shortages, energy crises, poverty, and biodiversity loss. As such, the DPRK government and the FAO have agreed that FLR should be prioritized to restore forests in the DPRK. This study aims to clarify (1) the current national efforts required to address deforestation and forest degradation in the DPRK; (2) the gaps; and (3) the major challenges in implementing FLR. Thus, this study aims to provide evidence in support of international communities formulating FLR strategies and action in the DPRK.

Analyzing Efforts for Afforestation and Forest Restoration in the DPRK

Deforestation and Forest Degradation

Approximately 1.5 million ha of forest were lost between 1979 and 2010; denuded forest land has rapidly increased, reaching 1,453,000 hm^2 in 2010 (Table 10.1). After 1995, deforestation and forest degradation became more serious. During the period 1990–2010, forest acreage decreased from 8.1 to 6.9 million ha, and the growing stock decreased from 526 to 330 million m^3, representing an annual reduction rate of 3%

Table 10.1 Changes in forest land by categories (Unit: 1000 hm^2)

Year	*1948*	*1954*	*1962*	*1979*	*1990*	*2000*	*2010*
Land with forests	7849	7557	8328	8435	8133	7554	6912
Denuded forest land	1007	–	473	338	432	870	1453
Non-forested land	358	–	214	197	380	405	352
Total	9214	–	9015	8970	8945	8829	8717

Table 10.2 Forest growing stock in the DPRK

Year	*1948*	*1962*	*1979*	*1990*	*1996*	*2000*	*2010*
Stocking Volume (10^6 m^3)	600	466	515	526	499	427	330

(Table 10.2). The DPRK is thus experiencing serious deforestation and forest degradation.

Deforestation and forest degradation are a critical concern for the DPRK and the international community. Afforestation and reforestation have been regarded as the main approaches for forest rehabilitation since the establishment of the DPRK (MoLEP, 2010); however, this approach has had little impact on reversing the trend of deforestation and forest degradation. In 2014, the DPRK considered serious actions, with the formulation of ambitious goals, including a 10-year National Forest Restoration Plan and Agroforestry Strategy and Action Plan with the assistance of the international community. These laid the foundation for future FLR.

National Forest Restoration Plan

Under the National Forest Development Perspective Strategy (2015–2044), a 10-year National Forest Restoration Plan (NFRP) was developed. The goal of this 10-year NFRP is to restore 1.68 million ha of deforested and degraded forest lands.

The course of actions is: first, afforestation and forest protection should be considered simultaneously. The major activities for afforestation include: (1) provision of sufficient seedlings; (2) planting trees in a responsible manner; and (3) cultivation of trees in line with scientific and technical requirements. Protection measures include: (1) prohibition of indiscriminative deforestation; (2) pest management; (3) prevention of forest fires; and (4) essentially resolving firewood issues. Second, forest sciences and technologies should be enhanced and applied, to (1) tackle challenges in breeding, planting, protecting, and managing forests; (2) introduce and disseminate advanced technologies for plantation and protection; and (3) promote the capacity of the Academy of Forest Sciences. Third, the control and supervision of plantation and protection of forests should be strengthened, with course of action including (1) control and supervision in a unified manner; (2) strengthened legal control on plantation and protection; (3) strict discrimination between the different management techniques; and (4) enhanced capacity of forest rangers and promotion of their roles and responsibilities.

To do this, efforts have been made to enhance tree seedling production capacities/nursery capacities to provide sufficient quantities of high quality tree planting materials required for the re/afforestation campaigns. The NFRP also considers that a maximum of 20% of areas should be restored through natural regeneration efforts. The plans intend to restore deforested and degraded forest lands not only through reforestation and afforestation but also through the introduction and widespread promotion of agroforestry on sloping lands, in addition to various protective measures including forest fire control and pest management.

For the implementation of the NFRP, the institutional arrangement from central to local levels needs to involve all relevant sectors including MoLEP, the Ministry of Agriculture, the Ministry of Timber Production, and the Ministry of Public Health. MoLEP has a key role and responsibility in planning and implementing the NFRP, by organizing the participation of all relevant stakeholder ministries, institutions, and agencies.

Agroforestry Strategy and Action Plan

With outside support from the FAO and Swiss Development Agency, agroforestry was recognized as a national strategy for livelihood improvement and environmental protection. Recognizing the advantages of agroforestry, in 2013, the DPRK declared that agroforestry should be vigorously promoted as a key strategy to restore forests and degraded lands in the country. The "National Agroforestry Strategy and Action

Plan (2015–2024)" (NASAP) was developed in 2015 to implement this policy.

The general goal of the National Agroforestry Strategy is "to contribute to the construction of a socialist power by improving food security and people's livelihoods, while restoring the country's landscape and environment through the active implementation of agroforestry in all provinces, cities, and counties." For the implementation of the national agroforestry strategy, the Action Plan identified the following key subject areas of action (specific goals):

1. Adopt agroforestry in non-forested and low productive forest lands to achieve food security and create a reliable base for the production of various raw materials while accelerating forest ecosystem restoration.
2. Implement agroforestry on sloping agricultural lands, improve forest cover, and create a sustainable and productive environment while maintaining food security.
3. Reduce disaster risks as much as possible, protect the overall environment, ensure a healthy environment for agricultural production, and reduce dependence on forests, by planting lots of trees in disaster-prone areas such as sloping arable lands and riparian lands.
4. Raise public agroforestry awareness and capacity, and make agroforestry extension the work of the entire society.
5. Constantly improve economic profitability and the effectiveness of agroforestry in environmental protection, and strengthen the scientific basis by conducting more research on agroforestry.
6. Strengthen institutional capacity and provide a legal and administrative basis for the development of agroforestry, by strengthening coordination and collaboration between relevant stakeholders and improving relevant laws and regulations.

NASAP also clarifies how the strategy and action plan will be implemented: "NASAP will be implemented by actively participating and cooperating with the relevant stakeholders such as MoLEP and other ministries and governmental agencies, local governments, and practical end users, as well as scientific research institutions, under the guidance of the Cabinet. All central and local governments will develop their own periodic and annual plans and implement them in the context of the national economic plan. Land users in the field, such as enterprises, cooperative entities, and local people, will make an annual land use plan to plant trees and protect forest as required by the national afforestation plan and also produce food and generate income. Every year, the implementation of the NASAP will be evaluated from the field level to central level, through which proper measures will be taken to share good experiences and learn from lessons and mistakes." At present, organizational and institutional arrangements have been made to implement the NASAP.

Status and Gap Analysis

Regarding FLR as an intersectoral and comprehensive approach, Sabogal et al. (2015) proposed multiple action areas for FLR planning and implementation in a country. These include the assessment of landscape degradation and restoration opportunities; enabling environmental policies, regulations, and laws; institutional settings; governance issues; technologies and approaches; private sector investment; resource mobilization; capacity development, extension, and dissemination; and research needs (Sabogal et al., 2015). Table 10.3 provides a primary assessment of the current status and gaps for each action area except for private sector investments. The private sector is extremely weak and supplementary to economic and social life in the DPRK and under control of the political regime. Through FLR training, a task force was developed involving participants from various sectors and various technical backgrounds, including officials from MoLEP, the Ministry of Timber Production, and the Ministry of Agriculture, and experts from the Academy of Forestry, Academy of Agriculture, Academy of Social Sciences, Academy of Sciences and Kim Il-sung University. By organizing participatory workshops, we established the following gaps that need resolving in order to implement FLR action in the DPRK.

- Extremely poor data access for outsiders, including UN agencies like the FAO. As the DPRK was identified as one of the priority states for the FLRM facilitated by the FAO, it has an obligation to collaborate with the FAO and other international organizations to provide relevant data for FLR; otherwise, it is unlikely that technical assistance and financial support will be provided to implement FLR.

Table 10.3 Current status and gaps categorized in key areas of intervention in DPRK FLR planning and implementation

Key areas of interventions	*Current status*	*Gaps*
Assessment of landscape degradation and restoration opportunities	1. Used GIS technology to assess land use change nationwide but lack capacity to produce precise small watershed mapping 2. Identified degraded lands based on sloping grades 3. Assessed agroforestry as successful restoration efforts 4. Primarily assessed linkage CBD and UNCCC with forest restoration 5. Identified the drivers to a certain extent; food shortage, shortage of rural energy/fuel, unsustainable land use practices, weak institutional capacities, natural disaster as a result of climate change	1. The assessment is not comprehensive and does not consider all factors like socioeconomic aspects 2. There are no experiences of using appropriate methodologies such as ROAM and LADA 3. Required comprehensive assessment of integrated watershed management 4. Limited knowledge on ecosystem approach, watershed mg. approach, REDD+ 5. No participation of stakeholders in assessment
Enabling environment	1. Dominated by sectoral oriented and fragmented policies, laws, and regulations to support FLR 2. Given afforestation and forest restoration as one of the national priorities 3. Revised policies, laws, and regulations quite often	1. Do not yet analyze policies, regulations across different sectors 2. Poor capacity to conduct analysis

(continued)

Table 10.3 (continued)

Key areas of interventions	*Current status*	*Gaps*
Institutional setting	1. Identified relevant land use sectors and stakeholders for FLR (forestry, agriculture, livestock/rangeland, energy, mining, etc.) 2. Formulated 10-year National Forest Restoration Planning 3. Formulated 10-year Agroforestry Development Strategy and Action Planning 4. Food security and livelihood improvement would be considered as an entry point in forest restoration plans 5. The current existing mechanism such as the Central Forest Restoration Campaign Committee can play a role in coordinating different sectors/stakeholders to engage in dialog. This committee organizes semi-annual progress review meetings and, if necessary, other multi-sectoral meetings 6. There is also the national Association for Promotion of Agroforestry Technology that organizes a scientific and technical forum on a regular basis	1. No mechanism to treat stakeholders as they are at stake, in particular communities and lay people 2. No legally binding national planning for CBD and UNCCC 3. Weak institutional support to level off national strategy and action plan between forest restoration and CBC, UNCCC, and UNCCD 4. Livelihoods lost due to political struggle 5. Approaches not ready for activities to involve multi-sectors and multi-institutions/institutions 6. MoLEP with SDC and EU 4, WFP. Partnership being developed, but requires strengthening

Key areas of interventions	*Current status*	*Gaps*
Governance issues	1. The land is state-owned, but usage rights are distributed to the various land use sectors and organizations 2. Centralized governance with prevailing top-down campaigns 3. Assess land-tenure issues and try to secure tenure, especially for local stakeholders, as a key issue to allow investments in FLR 4. Top-down decision-making process relying on experts and officials in senior levels 5. No mechanisms to facilitate engagement with all relevant stakeholder groups	1. No robust land tenure to user groups, which considered fundamental issues required to be addressed in agroforestry development in sloping land 2. Land use conflict occurred but little information disclosed 3. No knowledge available in the international concept of people's participation 4. There is great potential to improve decision-making processes as the DPRK is very unique internationally regarding decision-making processes 5. Presence of stakeholders at plantations is not an issue, but engagement is an issue. There is little room to discuss FLR in future
Technologies and approaches	1. Traditional and conventional methods of forest and landscape management were neglected, and more efforts went to introducing innovative and more sophisticated technologies and approaches 2. There are efforts to build on and disseminate the successful experiences and practices that already exist in Agroforestry, Conservation Agriculture, and Organic Farming 3. FLR is very new to the DPRK and requires more capacity building and FLR campaigns	1. Require knowledge sharing of successful experiences and approaches already carried out, including traditional knowledge 2. Identify knowledge gaps

(continued)

Table 10.3 (continued)

Key areas of interventions	*Current status*	*Gaps*
Capacity development and extension	1. Identified capacity development needs at the individual and organizational level in the short term 2. There are no capacity development programs. The existing sectoral and extension programs could partly contribute to FLR efforts 3. The DPRK promotes scientific and technical extension through the establishment of networks in all potential organizations and agencies. There are limited available sources on FLR 4. Limited national funding and support for national networks and extension services 5. The internal information service networks have improved however still far behind many developing countries	• Capacity development needs and strategies have been identified by relevant sectors, but not fully considered by all sectors, synergy issues,and thematic subjects • Multi-sectoral and multi-disciplinary aspects have not yet been considered in identifying capacity development needs and strategies • Need to develop networks/knowledge platforms between practitioners and extension services in order to disseminate good practices • Support the establishment and continued capacity strengthening of networks of practitioners and extension services
Resource mobilization	1. Limitations exist when trying to access donors (multilateral, bilateral, foundations, etc.) due to financial sanctions by the international community 2. Limitations exist when trying to access innovative sources of funding through mechanisms such as climate finance instruments and/or payment for environmental services, and when developing incentive packages that include economic and non-economic benefits 3. Extremely poor private sector in the DPRK 4. The state budgeting system is followed by the integration of the FLR action plan to the national and sectoral plans to streamline national investment	1. Desire to devise a coordinated approach to informing/sensitizing SDC, EU 4, and WFP and support the development of new project proposals in the DPRK 2. Required to integrate FLR into state budget and public investment funds 3. Required to develop a unique resource mobilization plan for the DPRK, as it is a unique state

Key areas of interventions	*Current status*	*Gaps*
Information dissemination and research needs	1. Current research aims to narrow the gap in advanced technologies between the DPRK and other states, however the gap has increased in 3 decades 2. Limited information sources on FLR	1. Required to identify (biophysical and socioeconomic) gaps in knowledge that research institutions could address more effectively 2. Emphasize research geared to innovative solutions for local stakeholders 3. Required to develop FLR relevant information with practical knowledge and experiences targeting varied audiences 4. Required to develop robust indicators adapted to the local/national context and develop consistent monitoring systems in order to improve the effectiveness of FLR efforts

- Extremely poor access to FLR knowledge and information. FLR is a new term to anyone in the DPRK. The DPRK has a closed information system that is disconnected from the growth in information and data in the rest of the world. Therefore, it is unlikely to get access to information regarding FLR on websites. Professionals rely on books or journal articles collected on the off chance from participating in international conferences and information provided by professionals visiting the DPRK.
- Neglecting traditions and experiences learnt, while expecting miracles to happen. In the counties I visited, I was expected to provide some kind of magic tree species that could survive without water and grow fast to produce sufficient fuelwood for residents. Influenced by traditional Chinese culture, the landscape was regulated through "integrated water and mountain management." Village forests play a similar role to fengshui forests in many places in China. Village regulation can be effective at managing community base forests. Professionals and officials in the DPRK began to consider FLR as a sort of new theory or measure to solve the problem of deforestation and forest degradation. After explaining that they should pay more attention to these traditions in future FLR action, they felt disappointed. However, some professionals have ample field experiences to highlight the importance of FLR in the DPRK.
- Poor coordination among various ministries and between central and local governments. I interviewed an expert who had lived in the DPRK for over a year while implementing integrated rural development projects, who said: "*no coordination among ministries; unlikely coordination among departments in a ministry; no talk at all between central and local authorities.*" However, participants from various institutions in the DPRK suggested that poor coordination was attributed to insufficient resources available for coordination requirements, such as travel costs and utilities.

Roadmap and Challenges

Despite being affected by economic sanctions from the UN with regards to reforestation and forest restoration, the FAO, alongside many donors, has assisted the DPRK in developing a primary roadmap to instill international FLR norms in the DPRK, to reverse forest cover decline. The DPRK was accepted as a primary beneficiary for the FAO FLRM, and the FAO provided 1 year of assistance in the readiness phase to develop the

DPRK FLR action plan. The action plan provides a platform for developing FLR synergy among international organizations and the DPRK regarding capacity building, piloting, campaigning, scaling up, good governance, and assessment. "Forest and landscape restoration for environmental protection and improved rural livelihoods in the Democratic People's Republic of Korea" has been approved by the FAO, which covers forest restoration in the low land areas in the DPRK. The Swiss Development Agency has 10 years of experience in sloping-land management in the DPRK and intends to scale up from agroforestry development at the site level to watershed or landscape levels, targeting hilly areas in particular for sloping-land rehabilitation and forest restoration. The World Food Programme provides substantial food aid to the DPRK and engages in a scheme called food for work, which might be able to provide assistance to the action plan of the FAO FLRM to develop "FLR Learn Center DPRK."

A phased project over 3 years will be followed for implementing the action plan, which aims to build up the capacity and develop "learning centers" established at selected counties that represent various geographic conditions and cover the most critical issues related to forest and landscape restoration for disseminating the technical know-how of integrated approaches and multi-sectoral coordination and multi-stakeholder collaboration mechanisms. Finally, the national FLR strategy will be developed based on the experiences generated from implementing the action plan. This will assist the implementation of the 10-year National Forest Restoration Planning and Agroforestry Strategy and Action Plan to restore 1.5 million ha of forests in the DPRK by the end of 2025.

FLR has many challenges to overcome in the DPRK that are extremely complex and multi-dimensional, as shown in Table 10.3. However, the following are the most important, as agreed by the interviewed experts:

- Change in behavior and attitudes of officials and technicians. The behavior and attitude of outsiders (policy makers and program officers) determine their relations with the local people. The outsiders need to recognize "the knowledge and skills of the community people," and to collaborate with the local people, to be "learners" rather than "leaders" or "alms givers."
- The public should participate in decision-making for forestry projects and policies at every stage of the project implementation cycle.
- Staff capacity needs improving, as it is low at nearly all levels, and local forestry agencies have gotten used to operating with top-down decision-making, despite the need for a bottom-up approach.

- The system of forest land ownership remains to be resolved, as the land tenure for user groups is insecure and of short duration. Connected pieces of land in one watershed are currently managed by various agencies reporting to various ministries.
- Forestry goals are typically targeted for conservation purposes or material production, and achieving a sustainable balance between the two has proven difficult. Governance in the DPRK has not yet been able to implement sustainable forestry adequately while balancing short- and long-term development goals.
- There is a lack of comprehensive and integrated approaches to address forest and landscape issues. Sectoral-oriented policies and legislative frameworks are dominant with no communication among various sectors.

Lessons Learned

There is need for a profound shift in attitudes toward the role of local people, recognizing that local people can be the main actors and partners in forest and landscape management. Reforms are needed in the forestry sector, and forest policy needs to be developed to support the transformation of local livelihoods to a forest-friendly paradigm that can escape the vicious cycle of poverty, hunger, deforestation, and environmental degradation. Future FLR policies need to focus on the decentralization of forest management, the clarification of forest tenure, and the improvement of multi-sectoral cooperation and multi-level collaboration. The role of traditional knowledge and customary regulations should be highlighted and forest management needs need to be integrated into socioeconomic development planning programs. China has similar experiences concerning the restoration of degraded forests, and it could be used as a model (Liu & Innes, 2015).

Local communities have been blamed for deforestation and forest degradation, due to use of large amounts of fuelwood required for cooking and heating, and slash-and-burn cultivation of food. Usually, it is the rural people and local communities that live near forests that are considered forest enemies.

There is a lack of seedlings, pesticides, and fertilizers as a consequence of international sanctions. Taking seedlings as an example, with the assistance of the Republic of Korea, the Central Nursery Station was established, and with the DPRK leader's call this center has expanded. However, a lack of seedlings is not a likely root cause of deforestation and forest degradation.

Lack of forest sciences and technology is another excuse for the reforestation and forest restoration failures. The Central Nursery Station demonstrated that the DPRK has access to and ability to use advanced technologies and facilities for producing seedlings. Agroforestry has gained legitimacy by being promoted as an innovative advanced technology for land use; however, agroforestry practices are similar to traditional knowledge. For instance, slash-and-burn is a type of agroforestry practice, which can allow the use of land in a sustainable way if the rotation period is sufficiently long (Luo et al., 2015).

Forests are arenas of struggle and conflict, where both trees and local forest dwellers usually find themselves on the losing side (Doornbos et al., 2000; Liu, 2006). Illegal farming on sloping land by user groups has contributed greatly to the food security issue in rural DPRK since 2000, which can be considered as unofficial decentralization of land management. Land use licenses issued to user groups could be legalized to provide land use rights to these groups; however, the current legal approach to centralize land use planning has caused land use conflicts between users and the public sector. This requires new policies and institutional arrangements at different levels of government, local communities, and the private sector in addition to financial delivery mechanisms, appropriate extension services, effective monitoring, and evaluation (FAO, 2006; Liu & Innes, 2015).

References

Bollenbacher, B. L., Graham, K. T., & Reynolds, K. M. (2014). Regional forest landscape restoration priorities: Integrating historical conditions and an uncertain future in the northern rocky mountain. *Journal of Forestry, 112*(5), 474–483.

Doornbos, M., Saith, A., & White, B. (2000). *Forests: Nature, people and power* (pp. 1–3). Blackwell Publishers Ltd.

FAO. (2006). *Global forest resources assessment 2005* (FAO Forestry Paper No. 147, pp. 28–30).

FAO. (2010). *Global forest resource assessment report*.

FAO/RECOFTC. (2016). *Forest landscape restoration in Asia-Pacific forests* (S. Appanah, Ed). FAO/RECOFTC

GPFLR. (n.d.). *Global partnership on forest landscape restoration website*. http://www.forestlandscaperestoration.org/

IISD. (2002, February 27–28). Summary of the international expert meeting on forest landscape restoration. *Sustainable Developments, 71*(1) (2 March 2002). http://www.iisdca/crs/sdcfr/sdvol71num1.html. Accessed 3 March 2015.

Laurance, W. F., Sayer, J., & Cassman, K. G. (2014). Agricultural expansion and its impacts on tropical nature. *Trends in Ecology & Evolution, 29*, 107–116.

Liu, J. (2006). *Forests in the mist—Livelihoods and responses to the natural forest protection program in China* (Wageningen PhD Dissertation, p. 87).

Liu, J., & Innes, J. (2015). Participatory forest management in China: Key challenges and ways forward. *International Forestry Review, 17*(4), 477–484.

Luo, Y., Liu, J., Zhang, D., & Dong, J. (2015). Actor, customary regulation and case study of collective forest tenure reform intervention in China. *Small-Scale Forestry, 14*(2), 155–169. https://doi.org/10.1007/s11842-014-9279-1

MoLEP (Ministry of Land and Environmental Protection). (2010). *Report of national forest resources inventory.*

MoLEP. (2011). *DPRK 4th report on CBD.*

Reed, J., Van Vianen, J., Deakin, E. L., Barlow, J., & Sunderland, T. (2015). Integrated landscape approaches to managing social and environmental issues in the tropics: Learning from the past to guide the future. *Global Change Biology, 22*, 2540–2554.

Sabogal, C., Besacier, C., & McGuire, D. (2015). Forest and landscape restoration: Concepts, approaches and challenges for implementation. *Unasylva, 245, 66*(3), 3–10.

Sayer, J., Sunderland, T., Ghazoul, J., Pfund, J. L., Sheil, D., Meijaard, E., Venter, M., Boedhihartono, A. K., Day, M., Garcia, C., Von Oosten, C., & Buck, L. E. (2013). Ten principles for a landscape approach to reconciling agriculture, conservation, and other competing land uses. *Proceedings of the National Academy of Sciences of the United States of America, 110*, 8349–8356.

Sayer, J., Margules, C., Boedhihartono, A. K., Dale, A., Sunderland, T., Supriatna, J., & Saryanthi, R. (2014). Landscape approache: What are the pre-conditions for success? *Sustainability Science, 10*, 345–355.

Scherr, S., Shames, S. A., & Friedman, R. (2013). *Defining integrated landscape management for policy makers* (Ecoagriculture Policy Focus No. 10). Ecoagriculture Partners.

Tscharntke, T., Tylianakis, J. M., Rand, T. A., Didham, R. K., Fahrig, L., Batáry, P., Bengtsson, J., Clough, Y., Crist, T. O., Dormann, C. F., Ewers, R. M., Fründ, J., Holt, R. D., Holzschuh, A., Klein, A. M., Kleijn, D., Kremen, C., Landis, D. A., Laurance, W., Lindenmayer, D., ...& Westphal, C. (2012). Landscape moderation of biodiversity patterns and processes—Eight hypotheses. *Biological Reviews, 87*, 661–685.

Van Oosten, C. V. (2013). Restoring landscapes—Governing place: A learning approach to forest landscape restoration. *Journal of Sustainable Forestry, 32*(7), 659–676.

West, P. C., Gerber, J. S., Engstrom, P. M., Mueller, N. D., Brauman, K. A., Carlson, K. M., Cassidy, E. S., Johnston, M., MacDonald, G. K., Ray, D. K., & Siebert, S. (2014). Leverage points for improving global food security and the environment. *Science, 345*, 325–328.

CHAPTER 11

Forests, Energy, Food, and Politics

DPRK Forests Are in Decline

The amount of forest coverage in the DPRK decreased during the period of Japanese colonization, increased slightly after liberation, decreased again during the Korean War, subsequently increased because of socialist economic progress, and then dramatically decreased in 1990, when Eastern European socialism collapsed. Approximately 1.5 million ha of forests have been lost from 1979 to 2010; denuded forest land has rapidly increased, reaching 1,453,000 hm^2 in 2010. After 1995, deforestation and forest degradation became more serious. Between 1990 and 2010, the amount of forest acreage in the DPRK decreased by 2.2 million ha from 8.1 to 6.9 million ha, and the growing stock decreased from 526 to 330 million m^3, representing an annual reduction rate of 3%. The DPRK is thus experiencing serious deforestation and forest degradation.

Grassroots efforts have been blamed for this deforestation and forest degradation. For instance, large amounts of fuelwood were harvested for cooking and heating while slash-and-burn cultivation was used for food crops. It is usually the rural communities living in forests that are blamed for any associated problems.

Deforestation and related social erosion have endangered the lives of North Koreans in flooding season. In 2016, floods killed hundreds, and many lost their homes (Jozuka, 2016). In the summer of 2018, for instance, severe flooding and landslides resulted in 76 casualties in

L. Jinlong, *Crying Forests*,
https://doi.org/10.1007/978-981-16-9725-8_11

Hwanghae province, many of whom were children. In response, the DPRK has attempted to implement sustainable land management policies (Nam, 2004).

Flooding caused by deforestation and forest degradation has also led to food shortages (FAO 2019; FAO/WFP, 2019). Flooding destroys not only buildings but also croplands, which can exacerbate famines. According to the United Nations (UN), which visited the DPRK after the 1995 and 1996 floods, there was a loss of 1.9 million tons of crops in addition to property damage. The annual recurrence of floods has exacerbated the problem of chronic starvation. As of 2017, approximately 41% of North Koreans are suffering from malnourishment (UNDP, 2018).

Frequent forest landslides often damage social infrastructure, such as roads and factories, creating new problems. From 1990 to 1998, coal production decreased from 33.15 million tons to 18.6 million tons; power generation decreased from 22.7 billion kilowatts to 17 billion kilowatts; steel production decreased from 3.36 million tons to 945,000 tons; crude oil imports decreased from 2.52 million tons to 500,000 tons. Due to serious shortages of raw materials, fuels, and electricity, most large and medium-sized enterprises are in a state of suspension or semi-stop production, and the operation rate of enterprise equipment is only 20%. In 2000, chemical fertilizer production decreased to almost none, and cement production decreased to below 2 million tons. After 2000, major industrial production slightly increased.

Deforestation and forest degradation have become a severe threat to food security, which may become an important factor in maintaining political stability in the DPRK. Environmental problems are closely connected with political stability in the DPRK. Food security continues to be a problem in the DPRK, which may lead to political destabilization in future (Habib, 2010).

DPRK government officials blame the lack of seedlings, pesticides, and fertilizers (a result of international sanctions) for their failure to reverse these problems. From an international perspective, however, corruption and poor governance are responsible for the major levels of deforestation and forest degradation seen in the DPRK (Liu et al., 2016). It is not known how current international politics or domestic development strategies and governance regimes will affect these forest cover changes. As such, the DPRK may serve as a case study for documenting lessons and learning from experience, although it is extremely painful for the people who live on the Korean Peninsula.

Juche Ideology Dominates Approaches to the Environment and Development

Juche Ideology is a constitutional norm, and defined as a people-centered worldview and a revolutionary ideology for achieving independence of the masses, or self-reliance—it is the DPRK's official governing principle for all aspects of its affairs and policies, including forest issues (Lee, 2003). Juche is an essential concept for understanding the DPRK's interpretation of and approach to the environment (Winstanley-Chesters, 2015). With its focus on the self-reliance of the North Korean people, the idea of Juche puts emphasis on the need to protect the natural environment for human well-being with a hygienic environment and working conditions. For decades, the DPRK has expressed a genuine desire for sustainable management of its natural resources (especially the land) for agricultural self-sufficiency. In that sense, Juche ideology has played a positive role in promoting the sustainable management of natural resources.

Self-sufficiency in food production has always been a national policy priority. The land was considered as the basis for the livelihood and prosperity of the people, and the government tried to maintain its fertility. When agricultural self-sufficiency was challenged in the 1970s due largely to scarcity of arable land, rural people were mobilized to cultivate terraces on mountain slopes less than 500 m above sea level, with a gradient of less than 15 degrees. The DPRK's forestry policy has focused on the use of forest resources for socialist economic development for a long time. The excessive use of forests for the past 60 years has severely impeded the natural rehabilitation of degraded forests. The people-centered idea of Juche has justified the conquest of nature, causing environmental degradation.

As a socialist country, the DPRK has suggested that pollution mostly occurs in capitalist systems where people and capital are supposedly driven by profits (Nam, 2004). In the DPRK, protecting the environment is an important work that shall be a permanent undertaking in building socialism and communism. Understanding and resolving these environmental problems is needed to improve local living standards.

Forests Matter to Food and Energy

Triggered by a combination of the withdrawal of favorable trade conditions with the East European group, particularly the former Soviet Union, and inefficient farming resources, the conversion of forested sloping land to food production was inevitable when faced with famine, which lead to widespread soil erosion and landslides. Meanwhile, the DPRK faces extreme weather conditions, often experiencing a few typhoons each year in the short rainy seasons during the summer months, which has led to flash floods, less rainfall in other seasons, and drought. All of these conditions have compromised the DPRK's ability to supply food for their people.

> Statistics on DPRK grain output began in 1961, when it was 3.58 million tons, and hovered between 3 and 4 million tons thereafter. From 1969 to the early 1970s, it exceeded 4 million tons. In the 1980s, it rose again to between 5 and 6 million tons. In the early 1990s, it broke 9 million tons, a record high (9.31 million tons in 1993). In 2010, grain output (such as rice and corn) was 4.52 million tons, which was similar to levels in the 1970s. The DPRK's food production in 2018 was the lowest level since 2008 at only 4.951 million tons, a decrease of 503,000 tons from 2017. In 2020, total grain output this year was 4.4 million tons, a decrease of 5.2% from the previous year.

The DPRK faced a serious political and economic crisis in the 1990s when the Eastern European socialist states collapsed. Development of its nuclear weapons program worsened its relationship with the United States in the 1990s and left it internationally isolated. The depressed economy was exacerbated by natural disasters that occurred in the mid-1990s, which greatly intensified its energy, food, and economic crises. As the economic crisis worsened, many North Koreans found ways to become self-reliant for food and fuelwood. They reclaimed sloped forests to produce their own food in times of widespread famine. Even public enterprises, organizations, and the army reclaimed forests to secure food and fuel supplies, and deforestation and forest degradation occurred naturally in the DPRK.

Wood has traditionally been used as fuel for cooking or heating and increased with population growth. Replaced by coal, petrol, gas, electricity, and biogas, fuelwood consumption reduced in the1970s and 1980s. As the DPRK has faced a chronic energy shortage since 1990, forests became the main energy source for households. People started cutting down trees to obtain firewood for cooking and heating; the amount of cut firewood increased from 3 million m^3 in 1990 to 7.2 million m^3 in 1996 (Park, 2010). But after the 1990s, fuelwood usage went up to 7,000,000 ~ 8,000,000 m^3 a year—a major cause of forest degradation in the DPRK. According to a census taken by UNFPA in 2008, timber was the main energy source for heating and cooking, proving that the North had a high level of dependence on forest products, especially in the countryside (UNFPA, 2008).

In the 1990s, the DPRK economy depended on imports and exports to and from China, and so the demand for timber soaked up. Specifically, timber was commonly exported to China, through both official and unofficial channels. Logging near the border between China and the DPRK was severe and used to gain access to foreign currencies, which was worsened forest degradation. Deforestation has weakened the role of trees in controlling droughts and floods and caused a vicious cycle that has perpetuated the food crisis in the DPRK.

Figure 11.1 shows electric power production from 1965 to 2015. The main sources of power in the DPRK are coal-fired thermal and hydroelectric power plants. Power production also decreased from a peak in 1990 and has been recovering since a low period in 1995. Rate of equipment utilization of coal-fired thermal and hydroelectric power plants decreased from a peak in 1990 and has slightly improved since 2000.

The DPRK has prioritized tackling energy and food shortages, as these are two major challenges currently facing the DPRK economy. Promoting the construction of small and medium-sized hydro-power plants contributed to an increase in the ratio of hydropower to coal-fuel generation, from 5:5 in 1980s, to nearly 6:4 in the 2000s.

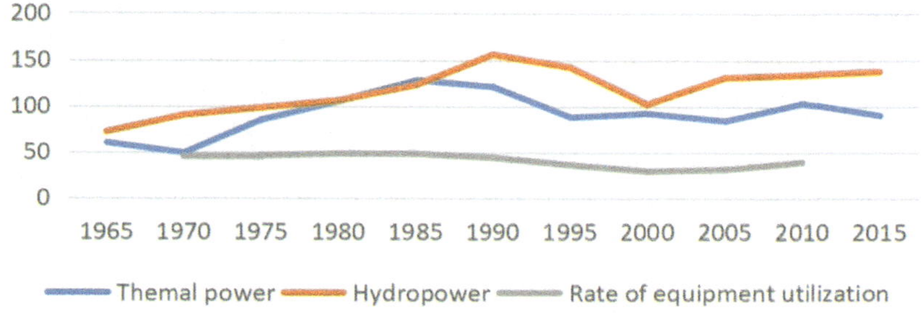

Fig. 11.1 Electric power production between 1965–2015 in DPRK (Unit: 10^8 KWH; %) (*Sources* www.kosis.kr)

Forests Matter to International Politics

Since the end of the Korean War, the history of the DPRK can be divided into two phases and the collapse of Eastern Europe was a key turning point. The DPRK demonstrated some successes in the first phase and failure in the second phase. Prior to the 1990s, the DPRK had achieved high levels of human development in terms of life expectancy, infant mortality, access to health services, water, and sanitation. The DPRK developed a highly planned economy and the government managed and allocated food, employment, housing, health care, and education. The country has been excluded from globalization, economic development, international trade, and investment system since this time. The DPRK is vulnerable to natural disasters and political tensions with outside countries, and minor shocks have serious effects on agricultural production and food availability, and so the DPRK has limited foreign currency with which to buy food internationally. The loss of socialist markets and natural disasters in the 1990s saw a reversal of many of the Human Development Index indicators: per capita income fell by 50%, life expectancy declined and the infant mortality rate increased (UNDP, 2011).

The DPRK is commonly perceived as one of the most secretive and uncooperative countries in the international community. In the field of sustainable development, however, the DPRK's cooperation with other states and international organizations has seemingly improved over time. While lacking in technical and financial capacity, the DPRK has joined a number of international environmental agreements and implemented various measures to fulfill its obligations as a contracting party (DPRK, 2012b).

Environmental issues, being a relatively neutral medium for dialogue, present some constructive opportunities for building trust between the DPRK and the rest of the world (Habib, 2013). There is a dilemma in promoting environmental protection in the DPRK for the purpose of establishing international peace and security. Environmental protection in developing countries like the DPRK requires assistance from the developed world, which is reluctant to provide aid unless the DPRK first gives up its nuclear program. It has always been an issue, and even now it seems there is no easy resolution.

The DPRK has shown willingness to collaborate with the international community on environmental issues (Kim & Ali, 2016). The

DPRK seems to have used environmental issues as channels of cooperative engagement with other countries and international organizations. Despite the imposition of the UN sanctions following nuclear tests, the DPRK has continued to send delegations to UNFCCC and has made statements emphasizing afforestation efforts. In 2016, former Minister of Foreign Affairs Lee Soo-yong delivered a speech at the signing ceremony of the "Paris Agreement" in the DPRK at the UN headquarters. He said: "The DPRK declared the restoration of mountain forests as a war on deforestation, and mobilized the whole country to vigorously carry out restoration of mountain forests. The DPRK plans to reforest 1.67 million hectares of mountainous land by 2024. The effect of mountain forests on absorbing greenhouse gases will increase the ability to reduce greenhouse gas emissions by more than 10 million tons per year. The DPRK believes that, in order to solve global environmental problems, priority should be given to ensuring world peace and stability." The afforestation plan is aimed at reducing greenhouse gas emissions, which contributes to achieving the DPRK's target—a reduction by 37.4% compared with the 1990 level (Guardian, 2015). Most recently, DPRK representatives highlighted how afforestation can contribute to environmental protection and economic development in the 2019 fourth session of the UN Environmental Assembly (UN Environmental Assembly, 2019).

South Korea Cares About Deforestation in the DPRK

Before 1970, the 38th Parallel was the frontier for the North to provide humanitarian assistance to the South, for the South to prevent revolutionary movements outside. After 2000, this line was named the "Division of Light." A satellite photo hung in the office of the former U.S Secretary of Defense Donald Rumsfeld when he headed the Pentagon. The photo showed the Korean Peninsula at night and depicted a luminous southern peninsula and dark northern half. The night view of the peninsula put electricity-strapped DPRK in stark contrast with the affluent South. During the day, the 38th Parallel line was the same color as the land that divided the two Koreas. Deforestation in the DPRK was third worst in the world. By contrast, the U.N. rated South Korea's forestation as the only successful case after World War II. In seasons of green foliage in the South, satellite images captured brown dirt land in the North.

Hungry DPRK residents dug up mountain foothills to make cultivation fields for food and even cut young trees for use as firewood in winter.

Many elder Koreans strongly believe that the Korean Peninsula is tiger shaped, and trees are seen as tiger hairs. Without hairs, tigers would die and so it is seen as unacceptable to pull down tiger hairs, as it represents the country's fortunes. The Baekdudaegan is a series of forested mountain ranges that runs through the majority of the Korean Peninsula, from Mount Baekdu in the north to Mount Jiri in the south. This mountain system has an important place in the spirit of the Korean people and in traditional pungsujiri (the Korean version of fengshui) philosophy and practices. The Baekdudaegan is a strategic choice that appeals to people of both the North and South (Na, 2012). The Republic of Korea (POK) National Assembly passed the Baekdudaegan Protection Act in 2003 to create a landscape-scale ecological corridor along the entire length of the mountain system in the South (Kim, 2015) and POK requested the DPRK consider extending the ecological corridor all the way to Mount Baekdu.

South Korea has successful experience in rehabilitating damaged forests, and the DPRK should learn from these successes. South Korea achieved a miracle in making its territory green with trees, something once seen in the north during large-scale forestation in the 1960s and 1970s. This experience along with financial support played a significant role in rehabilitating forests in the DPRK. South Korea has expressed their strong willingness to support the North's determination to rehabilitate its degraded forests (Hong, 1999). Indeed, both sides consistently included environmental cooperation as one of the ways to improve inter-Korean relations despite fluctuations in the relationship between the DPRK and outsiders.

The DPRK has managed to keep many of these ecological issues away from technology and politics, as did the South Koreans. During Park Geun-hye's administration, forestry cooperation with the DPRK was included in the scope of humanitarian assistance in order to promote non-governmental and local government exchanges and cooperation between the North and the South. South Korea proposed to start with small-scale environmental cooperation projects such as joint afforestation and exchange of seeds. The South Korean government also planned to assist the DPRK's reforestation and energy system, along with building an ecological belt throughout the Korean peninsula and tackling climate change. However, recognizing that the Park administration was conservative and considered the DPRK a primary enemy, the plan did not resonate with the DPRK, which dismissed the proposal quickly (Annie & Hasting, 2020). I once gave a report to the South Korean Congress and clearly pointed out that the DPRK forest issue is not just a food and energy issue, but also a political one.

It is not an issue of afforestation technology or sufficient seedlings in the DPRK. In October 2018, the DPRK and ROK convened to discuss detailed plans for forestry cooperation. There were four pledges requested from the DPRK on forestry: afforestation, pest control, exchange of technology on forest seedlings, and discussion on environmental challenges (Annie and Hasting, 2020). The Central Nursery Center was established with help from the ROK in the early 2000s, which can be expanded. However as demonstrated before, seedlings were not a root cause of deforestation and forest degradation in the DPRK. What can both sides do? Materials-based cooperation only. Due to political circumstances in the Korean Peninsula, there was no way to address institutional and political issues.

Great Efforts but Little Gain on Forests

In order to resolve the underlying problems of persistent starvation linked to frequent flooding, the DPRK has promoted afforestation policies at a domestic level. With the increasing severity of deforestation, the government enforced a regulation to prohibit further deforestation (Park et al., 2013). The first long-term policy plan was released in 2001, with the objective of cultivating 1.5 million hectares of forest from 2001 to 2010 and more ambitious plans were put forwards for the period 2011–2020. The central government established local agencies to implement the plan and banned deforestation for the use of cropping. However, the success of the plan is less clear.

Without guaranteeing peace and stability, there is no satisfactory solution for international cooperation, and any actions in the DPRK will be questioned. International donors welcomed a bottoms-up participatory process of developing locally appropriate agroforestry in the DPRK, helping to reverse the chronic food shortages and afforestation to retain land degradation. The projects supported by FAO and SIDA used innovative agroforestry technologies to provide food, fodder, and other products for local people while restoring degraded land. In the project sites, there were some positive responses, for instance establishing user groups, participatory tree species selection, participatory monitoring and evaluation, and presence from the international community; however, it hardly reached other regions due to the dominance of Juche ideology in society. At international aided project sites, sloping-land user groups—made up primarily of those most affected by the lack of access to the public food distribution system such as retirees and housewives—obtained rights-to-use, rights-to-harvest, and rights-to-plan or access to sloping lands for tree products and food. Outsiders saw it as a great achievement that these groups can be self-sufficient with technical support. However, the DPRK is not an individualistic society, but one we as outsiders may not understand, like many other developing countries in Africa and South Asia.

Forests are arenas of struggle and conflict in which both trees and local forest dwellers usually find themselves on the losing side (Doornbos et al., 2000; Liu, 2006). Illegal user-group farming on sloped land greatly contributed to food security in rural areas of the DPRK around the year 2000. These practices have also been used to improve recent food production rates, a circumstance that can be considered a product of unofficially decentralized land management. Issuing land use licenses to user groups was part of centralized land use planning, that brought land use conflicts between users and the government. Agroforestry in slope land could have prevailed in the DPRK if there was an introduction of new policies and institutional arrangements at different levels of government, local communities, and the private sector, as well as new financial delivery mechanisms, appropriate extension services, and effective monitoring and evaluation practices, as part of a bottom-up approach.

Since the 1970s, especially after the end of the Cold War, the world has entered a new pattern of peace and development as a theme of the times, and the third wave of economic globalization has emerged. Economic globalization with free trade at its core has contributed to great trade prosperity, great investment convenience, great flow of personnel, and great technological development. It has become an irreversible trend of the times in the world today. Driven by economic globalization, science and technology are advancing with each passing day, transportation, communication, and other infrastructure are developing rapidly, and the cost of circulation of goods and information on an international scale has drastically reduced; the in-depth development of finance has made international financing more convenient, and the scale of global capital flows is increasing. More countries and regions have opened up to the outside world, choosing to integrate into the world economic system and face the international market to develop their economy, which has greatly expanded the depth and breadth of the international division of labor. The DPRK is a special country, which is completely isolated. The DPRK has faced United Nations Security Council (UNSC) sanctions. While many countries embarked on a process of rapid globalization, economic transformation, and expansion of information technology, the DPRK suffered from isolation partly due to UNSC sanctions and partly due to Juche ideology.

In order to rehabilitate degraded forests, several fundamental problems must be addressed. Most importantly, the people of the DPRK must gain greater security through economic recovery. Only in this way will this high dependence on forests for food and firewood be resolved. Reversal of forest quantity and quality requires comprehensive measures, including favorable international relations, political and institutional reforms, the decentralization of forest management, and proper technical support for local communities. The forestry issue is complex and, therefore, no longer solely the prerogative of the forestry sector, but is embodied in political, social, cultural, economic, environmental, and other broad development issues. Without settling down all these complicated issues, forests in the DPRK undoubtedly become further degraded and deforested.

References

Doornbos, M., Saith, A., & White, B. (2000). *Forests: Nature, people and power* (pp. 1–3). Blackwell Publishers.

DPRK. (2012b). 2nd Communication on Climate Change.

FAO. (2019, May 3). *After worst harvest in ten years, 10 million people in The DPRK Face imminent food shortages*. http://www.fao.org/news/story/en/item/1192420/icode/

FAO & WFP. (2019). *FAO/WFP joint rapid food security assessment—Democratic people's republic of Korea, Bangkok*. http://www.fao.org/3/ca4447en/ca4447en.pdf

Guardian. (2015, December 8). The DPRK launches 'war on deforestation'. *The Guardian*. https://www.theguardian.com/

Habib, B. (2010). Climate change and regime perpetuation in The DPRK. *Asian Survey, 50*(2), 378–401.

Habib, B. (2013). *The DPRK's surprising status in the international climate change regime*. www.eastasiaforum.org/2013/11/09/north-koreas-surprising-status-in-the-international-climate-change-regime

Hong, S. J. (1999). Environmental pollution in The DPRK: Another South Korean burden? *East Asian Review, 11*, 79–98.

Jozuka, E. (2016, September, 30). The DPRK floods: Survivors homeless, sick as winter looms. *CNN*. https://edition.cnn.com.

Kim, R. E. (2015). *Legal brief on the Baekdudaegan protection act, 2003*. International Development Law Organization.

Kim, R. E., & Ali, S. H. (2016). Green diplomacy—An opportunity for peace-building? *Environmental Policy and Law, 46*(1), 86–96.

Kyung-Seok, P., Lee, S.-Y., & Park, S.-Y. (2013). Changes in The DPRK's forestry policy. *The DPRKn Review, 9*(1), 20–29.

Lee, G. (2003). The political philosophy of Juche. *Stanford Journal of East Asian Affairs, 3*, 105–112.

Liu, J. (2006). *Forests in the mist—Livelihoods and responses to the natural forest protection program in China* (Wageningen PhD Dissertation, p. 87).

Liu, J., Liang, M., Li, L., Long, H., & De Jong, W. (2016). Comparative study of the forest transition pathways of nine Asia-Pacific countries. *Forest Policy and Economics, 76*, 25–34. https://doi.org/10.1016/j.forpol.2016.03.007

Na, H. R. (2012). Nationalism as a factor for an international environmental regime: Korea and the East Asian Biosphere Reserve Network (EABRN). *East Asian Science, Technology and Society, 6*, 83–99.

Nam, S. (2004). The legal development of the environmental policy in the democratic people's republic of korea. *Fordham International Law Journal, 27*(4), 1322–1342.

Park, J. H. (2010). A research on the actual condition of The DPRK's forest devastation through MODIS satellite images. In *The north and south' s cooperation for the DPRK's forestation* (p. 7). Institute for Peace and Unification Studies at Seoul National University.

Song, A. Y., & Hastings, J. V. (2020). Engaging The DPRK: Environmental cooperation in peace building. *Third World Quarterly, 41*(11), 1809–1927.

UNDP. (2011). *Country programme for the democratic people's Republic of Korea (2011–2015).*

UNDP. (2018). *DPRK needs and priorities*. https://www.undp.org/content/dam/unct/dprk/docs/unct_kp_NP2018.pdf.

UN Environmental Assembly. (2019). *National statement of democratic people's republic of Korea at the fourth session of the United Nations environment assembly*. http://web.unep.org/environmentassembly/democratic-peoples-republic-korea.

Unite Nations Population Fund (UNFPA). (2008). *DPRK population census national report*.

Winstanley-Chesters, R. (2015). *Environment, politics, and ideology in The DPRK: Landscape as a political project*. Rowman & Littlefield.

Postscript

Part I

Last Sunday, escorts from the Ministry of Land and Environmental Protection were informed that I wanted to go for an evening walk outside the hotel where I lived. A young guy—who graduated from Kim Il Sung University and worked for the Ministry of Land and Environmental Protection—put down everything on hand and walked out of the hotel with me, as usual. This time the boy's behavior was a little strange. He stayed within close proximity to me, a distance of about one meter. He smiled and always tried talking to me. The whole DPRK was filled with excitement. Late last night, it successfully launched the Hwasong-14, an intercontinental ballistic missile that can reach the western region of the United States. Instead of my usual arrangement of walking along the Daetong River, I accepted his recommendation and watched the singing and dancing in celebration of the missile launch. After turning south from the east side of the Haebeishan Hotel, and walking 500 m, we arrived at the Pyongyang Theatre. The front entrance of the theater turned into a stage for performances, and the outer square became the auditorium. At least 2,000 spectators turned up. By the time we had arrived, the performance was almost over, and there were only two shows left, the sketch and the chorus. With the help of the escort, I understood the content of the show. The main idea of the sketch was that three soldiers wounded in battle went to the river to clean their wounds, met a group of beautiful Korean girls who just happened to be cleansing themselves, and then…

L. Jinlong, *Crying Forests*,
https://doi.org/10.1007/978-981-16-9725-8

(I'm sure you can guess what happened next). The chorus was also a small group of actors and actresses, with the main idea here to use everything they had, blood and flesh, spirit and soul to defend the great leader. The performance ended with loud singing and excited cheers.

We turned around and headed north, and returned to the place where I would walk every day when I was in Pyongyang, the best place for a daily walk—the Daetong River near Kim II Sung Square. The escort recommended we head to Kim II Sung Square to watch square dancing. A full crowd of more than 10,000 men and women sang and danced in pairs in the western section of Kim II Sung Plaza. They were all girls and boys in their 20s, dressed in traditional Korean attire. They came from universities in Pyongyang and took their own transportation to Kim II Sung Square to dance. The guy who accompanied me on the walk was immersed in incomparable joy and proudly said: DPRK has succeeded, DPRK has won, and people all over the country are singing and dancing with delight. I asked him if he had come here to dance. He said sadly, "I'm not qualified, yet. My thoughts and willpower are not up to standard, and I am not qualified to dedicate myself to the cause of defending the head of state. I need to work harder to become great, to become a good fighter for the leader." I comforted him: "Even if I begged my son, he wouldn't come here to dance". He felt some comfort.

Part II

Since 2006, I have visited the DPRK more than seven times, and the total number of days has exceeded 125 days. The DPRK ranks fourth in terms of countries I have visited most often and total number of days abroad. However, unlike other countries I have visited, such as the Netherlands, Canada, the Philippines, and Thailand, the DPRK is one of the few countries that I have studied hard, including forests and forest-related history, environment and natural resources, society and culture, history and governance. However, every time I come to the DPRK, I feel that it is a very strange country.

This is a country that seems to be immersed in fanatical idealism. Next to a portrait of the late leader, there was a floor fan shaking its head and blowing a breeze to the portrait of the leader. The meeting room that I reported to only had a single charger to operate the desk fan, which is considered a courtesy to UN experts. On the Day of Victory, a national holiday to mark the signing of the Korean Armistice Agreement, men and women held plastic flowers in their hands and bowed to the portraits of leaders, as part of a solemn ceremony. When you live in it, it's more like a society full of secularism. Among the people leaving China at Capital Airport, I saw North Koreans wearing leaders badges who had purchased many cartons of Japanese cigarettes—some had more than 20 cartons. At Pyongyang airport, North Koreans and customs officers were talking and laughing. Naturally, they took out one of the cartons of cigarettes. The customs officers quickly accepted it, and, as if they hadn't seen anything, continued to pretend to check the luggage of the entry personnel. On arriving in the DPRK this time, officials from the Ministry of Land and Environmental Protection arranged a two-day visit to the countryside and attached a request to provide a reception fee of US$2,000 so that they could purchase gasoline to arrange for me to accompany them. I forwarded their request to the FAO Representative in the DPRK. According to provisions in the UN sanctions documents, the UN agencies cannot give any financial support to the relevant DPRK agencies. Accordingly, I requested the Ministry of Land and Environmental Protection to cancel the visits to the countryside. I was actually very worried about my safety. For more than 10 years, I had used the same car, a Toyota Land Cruiser 4500 assisted by the United Nations Development Agency. When I went to the countryside in the summer of 2016, I found that the car's shock absorption system was completely malfunctioning. When driving up a bridge, the engine stalled due to lack of power, so everyone had to get out of the car and push together. This visit to the DPRK coincided with the rainy season, so I was naturally more worried. However, they insisted that I go to the countryside for inspection. While traveling to the countryside there was a person lying in the trunk of the car. When they arrived at the county, apart from translating, shaking hands, and saying goodbye, they were left with a meal of wine and food. Lying in a trunk is uncomfortable, and the roads to the countryside are mostly gravel roads. In haste, the driver reached speeds of 60 km per hour. I was sitting in the co-pilot's position, and it was all very uncomfortable. I questioned his reasoning. Officials from central government organizations should not bend their waists just for a good meal. On a county visit, the vice chairman of the People's Committee happily introduced local specialties, fine wines, fresh fish, and beauties-a paradise for men. I said: South Korea is very willing to reunite with the North, because the DPRK has beautiful women. Everyone laughed.

This is a very orderly, organized country. At the bus stop next to the Haebangshan Hotel, everyone queued no matter how long the line was. In tree-planting season, all able men, women, and children planted trees on the mountains and plains with the only requirement being that they could climb the mountain. It's hard to imagine how they organized and distributed tasks. On Saturdays, every lawn was filled with women holding small scissors and cutting the grass, as carefully as if giving a child a haircut. It was amazing to see tens of thousands of people who had all used public transportation, gathered together, and dancing in Kim Il Sung Square. However, this is seemingly a very disorderly organized country. Once you leave the main road near Kim Il Sung Square and enter the community, a pungent smell of human dung fills the air, just like in the morning in the alley near the Colosseum, the smell was really unbearable. After heavy rain, the Daetong River would rise near the Kim Il Sung Square, and the yellow-black river tumbled with garbage of various colors.

This is a country where everything goes according to plan. In 2014, the top leader of the DPRK approved the 2015–2024 forest restoration plan. It planned to reforest 1.68 million hectares while strengthening forest protection. Leading institutions were established from the central to the local level, and it was planned that scientific research, seedling preparation, afforestation, fire avenues, firewood forests, forest rangers, and forest police will be allocated 1.68 million hectares for every afforestation plot each year. The party, the government, the military, and the people are all mobilized and implemented level by level. Each county put the ten-year plan into a planning map of each plot and hung it on the walls. Officials I met were convinced that their plan was perfect and would be implemented without compromise. However, I told them that the DPRK had formulated a 10-year afforestation plan from 2001 to 2010 and planned to add 1 million hectares of planted forests. The area of planted forests in the DPRK was about 1.4 million hectares in 2000, but had reduced to 1.2 million hectares by 2010. Officials from the Ministry of Land and Environmental Protection stammered and argued that there were many reasons for this, mainly caused by fire, floods, and pests.

This is a country that has never acted according to plan, a country that makes people wonder. In the spring of 2016, together with the Ministry of Land and Environmental protection, I made a detailed work plan for the formulation of a forest landscape restoration action plan for the DPRK. Every activity planned is guaranteed by FAO funds. It had been repeatedly confirmed by officials from the Ministry of Land and

Environmental Protection and had been confirmed through a seminar involving multiple ministries and international organizations. However, in the summer of 2017, when I visited the DPRK again, none of this plan had been implemented. The only thing of concern in the meeting was to ask FAO to honor the $2,000 on the budget list for the purchase of gasoline. I gave my counterpart a nickname—Agreed but no Action. The old man accepted it very happily, and his colleague thought the nickname was very appropriate.

Part III

Documents on the DPRK's 10-year forest restoration plan attached great importance to the role of science and technology in forest restoration, and the language used astonished me, even as a scientific and technological worker, as it seemed as though it was the home of scientists worldwide. The faculty apartment of Kim Il Sung University is one of the best resident areas in Pyongyang, and the newly built science and technology museum and science city are the best buildings in Pyongyang. Over the last few decades, China has generally paid little attention to scientific research, in particular, research on basic sciences, and only attached importance to a sort of skill set. Many of these so-called scientific and technological researchers who are able to please officials and solve technical and practical problems become popular and are promoted. However, the reality is that compared with the DPRK, science and technology researchers would be much happier living in China. The connotation of sciences and technology in the DPRK has also been completely subverted.

In 2007, I introduced two Chinese controlled spray seedling systems for the DPRK to improve the reproduction of high-quality seedlings, as they are difficult to root in the Central Nursery Station under management of the state authority. This system automatically sprays according to the soil moisture of the container matrix and air humidity, without covering, and is used to propagate seedlings of precious tree species that are difficult to cultivate with root cuttings. The Chinese Academy of Forestry has intellectual property protection for this system and applied for invention patents after winning the National Invention Award. To my surprise, after 2014, this system was copied and equipped in nurseries in every county. By 2017, every county nursery had at least two sets of automatic controlled spray nursery systems. However, none of the counties I visited were using this system properly. Some used it to breed willows that are easy to reproduce, and some used it to breed seedlings of larch. Most of them were covered with shade materials, and one of the most important characteristics of the system was the fact that it did not require any shading or plastic covers for maintaining soil moisture. The technicians and officials from the Ministry of Land and Environmental Protection who accompanied me on the visit were unwilling to discuss technical issues on site to properly operate this system. The technicians in the county nursery said that this system was good, but I didn't know why. They wanted even more, similar systems. To me, it seemed that technology did not even need to solve problems, as long as it sounded good in theory. The correct use of technology and the redevelopment of technology adapted to local conditions hadn't been considered.

By 1991, Eastern Europe had collapsed, the former Soviet Union's annual supply of around one million tons of oil was interrupted, and the energy issue became an important bottleneck for the DPRK's development. Large-scale deforestation has brought serious natural disasters such as floods and droughts. After the mid-1990s, the DPRK fell into a severe food crisis, which caused proliferation of slash-and-burn agriculture that had stopped for more than two decades. The opening-up of forest land used for growing grain was not counted as farmland, and the farmland formed part of the grain quota (10% of the output was given to primary and secondary schools, 40% of the output was given to the country, and the rest was distributed among members). Most members of the county forest management committees were from villages, and they also faced severe food shortages. Employees of county forest management committees often turned a blind eye to allow villagers to clear up forests and grow grain without paying any share of their harvested grain. As early as 2013, during the process of discussing legislation in the agroforestry law, the government tried to legalize privately developed land and required villagers to hand in 10% of the grain harvest. I strongly opposed this policy. I have provided a lot of training to relevant officials and continue to provide opportunities for MoLEP officials to visit China so that they can understand the reasons for my stance. In the end, this policy of levying taxes on agroforestry was not implemented. To grow grain in woodland, it is impossible for villagers to use chemical fertilizers and pesticides, as the yield is extremely low. In good weather, the output of grain is only 3–4 tons per hectare. If severe droughts and floods occur, there is often no harvest.

After 2000, the population of cities such as Pyongyang remained unchanged, and the urbanization rate of the DPRK generally declined. It is likely some residents were moved to mountainous areas, as an increasing number of mountainous areas were opened up as permanent farmland. According to my personal estimation, of the DPRK's 9 million hectares of forest land, only about 5 million hectares of forest remain. Of the other 4 million hectares, at least one quarter of this land was illegally cultivated to grow food. After 2010, agroforestry was accepted by North Korean leaders and became an important national strategy. The legal, technical, and management system to promote the development of agroforestry was quickly established. Although local people have had authorized use of the slopes for 3–7 years, the target is 300,000 hectares. But the question is, who planted the trees on the state-owned sloping land? Government tree planting restricts people's use of the land. When the trees grow up, they all belong to the government, and the farmers can harvest food grown under the forest. The survival rate of afforestation is often very low, generally only 20–30%. The possibility of afforestation is also extremely low. It is impossible to do anything with government agencies constantly disturbing the land used by farmers.

A certain area of agroforestry is allocated to each county. In Pongsan County, a total of 600 hectares of agroforestry was approved for development. Approximately 1,000 members of the land use team own it. The teams using the land are responsible for afforestation and the seedlings are purchased by members at the county nursery. Fast-growing tree species can only be intercropped for 5–7 years, while slower-growing tree species can be intercropped for 8–10 years; this model is only used in early intercropping in plantations. The Ministry of Land and Environmental Protection has reduced expenditures and achieved the goal of afforestation and greening, and it is more effective than the mobilized afforestation model, more so than the development of agroforestry that we envisioned. As a technological system, agroforestry has been alienated as a means to solve sectoral problems, rather than to solve environmental and social problems. This is completely different to my vision of promoting the development of agroforestry in the DPRK since 2006.

Part IV

Since 2013, I have come to the DPRK every year and stayed at the Haebangshan Hotel. There are only a few kinds of breakfast buffet, and lunch and dinner have even less choice. The menu is not long and has never changed; very few dishes in the menu are appealing to me. It's bearable for one or two days, but after a week, it's really unacceptable. It was hard for me to be given another choice of hotel. This hotel is located in the center of Pyongyang. It is the starting point of the road from the capital to all parts of the country. It is also close to the political center of the DPRK. There are quite a few high-profile officials and beautiful women who wear badges rushing in and out of the hotel. The waiters that serve food in the restaurant and the female traffic officers outside the hotel look heroic from one the side and particularly charming from the other. Even the women grooming the lawn and the elderly ladies doing morning exercises are well-proportioned and attractive. However, on observing dozens of women waiting in line at the bus station, I noticed one was shorter than the other, with smaller buttocks and thinner legs. What puzzled me was the non-uniform height of the high-heeled shoes. I visited markets for local consumption, and there was no trace of diet pills. In residential communities, women took care of their food. Potatoes were only as big as peas, eggs as big as quail eggs, and sweet potatoes were like lentils. At the Haebeishan Hotel, the rice was mixed with a few little stones. Fortunately, my teeth are good, otherwise I would have been injured long ago. I think most of them are malnourished!

One day, a group of technical consultants hired by the United Nations and environmental management project officials from various embassies chatted together. Climate change, Trump, the DPRK, and human extinction were the most interesting topics. The conclusion was very unified: Capitalism, this demon, has completely destroyed the nature on which mankind depends for survival. Human desires are completely crazy under the spell of capitalism. In 2200 AD, mankind perished on this planet. Professor Singh, an Indian graduate from the MIT Department of Economics who has since worked at the London School of Economics, added: People here have survived, and people in other places have perished because there is no capital here. In 2030, the DPRK will be completely independent and will completely eliminate all connections with other countries and regions on this planet in terms of personnel, goods, and information. After 2040, men in other parts of the earth will have gradually lost their reproductive ability, which eventually will lead to the demise of mankind. This Indian economist was clever and knowledgeable. The explanation given was: The desires of the people here are controlled by the main thought, and the temperance of desires maintains the abilities of men. It was also blessed because other places perished, and the people in this place survived.

Game Over! game Should Be Over

August 5, 2017, Pyongyang

References

Agrawal, A., Chhatre, A., & Hardin, R. (2008). Changing governance of the world's forests. *Science, 320*, 1460–1462.

Antuma, J., Esch, B., Hall, B., Munn, E., Sturges, F. (2014). *Restoring forests and communities: Lessons from the Collaborative Forest Landscape Restoration Program*. School of Natural Resource and Environment, University of Michigan.

Aronson, J., & Alexander, S. (2013). Ecosystem restoration is now a global priority: Time to roll up our sleeves. *Restoration Ecology, 21*(3), 293–296.

Bastin, J. F., Finegold, Y., Garcia, C., Mollicone, D., Rezende, M., Routh, D., Zohner, C. M., & Crowther, T. W. (2019). The global tree restoration potential. *Science, 365*, 76–79.

Besseau, P., Graham, S., & Christophersen, T. (Eds.). (2018). *Restoring forests and landscapes: The key to a sustainable future*. Global Partnership on Forest and Landscape Restoration.

Bollenbacher, B. L., Graham, K. T., & Reynolds, K. M. (2014). Regional forest landscape restoration priorities: Integrating historical conditions and an uncertain future in the Northern Rocky Mountain. *Journal of Forestry, 112*(5), 474–483.

Clement, F. (2010). Analysing decentralised natural resource governance: Proposition for a "politicised" institutional analysis and development framework. *Policy Sciences, 43*(2), 129–156.

Central Statistic Bureau. (2010). *Statistical report of land use*. Central Statistic Bureau (in Korean).

Central Statistic Bureau. (2011). *Statistical report on achievement of MDA in DPRK*.

L. Jinlong, *Crying Forests*,
https://doi.org/10.1007/978-981-16-9725-8

Central Statistic Bureau. (2008). *Statistical report of population in DPRK.*

Chazdon, L. (2008). Beyond deforestation: Restoring forests and ecosystem services on degraded lands. *Science, 320*, 1458–1460.

Chazdon, L., Brancalion, P., Laestadius, L., Bennett, A., Buckingham, K., Kumar, C., Moll-Rocek, J., Guimarães Vieira, I., Wilson, S. (2016). When is a forest a forest? Forest concepts and definitions in the era of forest and landscape restoration. *Ambio, 45*(5), 1–13.

Longshan, C. (2003). Analysis on the characteristics and trend of The DPRK's foreign policy [J]. *Northeast Asia Forum, 4*, 8–11. (in Chinese).

Choi, I. H., & Woo, J. C. (2007). Developmental process of forest policy in korea and present status of forest desolation in The DPRK. *Journal of Forest Science, 23*(1), 35–49.

Chun, Y. W., & Tak, K.-I. (2009). Songye, a traditional knowledge system for sustainable forest management in Choson Dynasty of Korea. *Forest Ecology and Management., 257*, 2022–2026.

De Jong, W., Galloway, G. (2020). Narratives on the sustainable development goals and tropical forests. In J. Blaser & P. Hardcastle (Eds.), *Achieving sustainable management of tropical forests*, in press. Burleigh Dodds Science Publishing.

DPRK. (2011). *4th report on biodiversity* (English)

DPRK. (2012a). *Environment and climate change outlook.*

DPRK. (2012b). *2nd Communication on climate change.*

DPRK. (2016). *Intended nationally determined contribution of democratic people's republic of Korea.*

Doornbos, M., Saith, A., & White, B. (2000). *Forests: Nature, people and power* (pp. 1–3). Blackwell Publishers.

Erbaugh, J., & Oldekop, J. (2018). Forest landscape restoration for livelihoods and well-being. *Current Opinion in Environmental Sustainability, 32*, 76–83.

FAO. (2006). *Global forest resources assessment 2005* (FAO Forestry Paper No. 147, pp. 28–30).

FAO. (2001). *How forests can reduce poverty*. FAO and DFID.

FAO. (2010). *Global forest resource assessment report.*

FAO. (2020). *Global forest resource assessment report.*

FAO. (2019, May 3). *After worst harvest in ten years, 10 million people in The DPRK Face imminent food shortages*. http://www.fao.org/news/story/en/item/1192420/icode/

FAO/RECOFTC. (2016). *Forest landscape restoration in Asia-Pacific forests* (S. Appanah, Ed). FAO/RECOFTC

FAO & WFP. (2019). *FAO/WFP joint rapid food security assessment—Democratic people's republic of Korea, Bangkok*. http://www.fao.org/3/ca4447en/ca4447en.pdf

Franzel, S, Hitimana, L,& Akyeampong, E. (1995). Farmer participation in on-station tree species selection for agroforestry: A case study from Burundi. *Experimental Agriculture, 31*(1), 27–38. Fao/Recoftc. (2016). *Forest landscape restoration in Asia-Pacific forests* (S. Appanah, Ed).

German, L. A., Kidane, B., & Shemdoe, R. (2006). Social and environmental trade-offs in tree species selection: A methodology for identifying niche incompatibilities in agroforestry. *Environment, Development and Sustainability, 8*(4), 535–552.

Godfray, H. C. J., Beddington, J. R., Crute, I. R., Haddad, L., Lawrence, D., Muir, J. F., ... & Toulmin, C. (2010). Food security: The challenge of feeding 9 billion people. *Science, 327*(5967), 812–818.

Görg, C. (2007). Landscape governance: The "politics of scale" and the "natural" conditions of places. *Geoforum, 38*(5), 954–966.

GPFLR. (n.d.). *Global partnership on forest landscape restoration website*. http://www.forestlandscaperestoration.org/

Guariguata, M, Brancalion, P. (2014). Current challenges and perspectives for governing forest restoration. *Forests, 5*(12), 3022–3030.

Guardian. (2015, December 8). The DPRK launches 'war on deforestation'. *The Guardian*. https://www.theguardian.com/

Jingli, G., Lei, W., & Kai, C. (2015). Current situation, problems and countermeasures of grain production and trade in The DPRK. *World Agriculture, 06*, 170–174. (in Chinese).

Gilmour, D. (2005). An historical account of fuelwood restoration efforts. In S. Mansourian, D. Vallauri & N. Dudley (Eds.), *Forest restoration in landscapes: beyond planting trees* (pp. 223–227). Springer.

Habib, B. (2010). Climate change and regime perpetuation in The DPRK. *Asian Survey, 50*(2), 378–401.

Habib, B. (2013). *The DPRK's surprising status in the international climate change regime*. www.eastasiaforum.org/2013/11/09/north-koreas-surprising-status-in-the-international-climate-change-regime

Habib, B. (2015). Balance of incentives: Why The DPRK interacts with the UN framework convention on climate change. *Pacific Affairs, 88*(1), 75–97.

Haggar, J., Ayala, A., & Dı´az B, Reyes CU. (2001). Participatory design of agroforestry systems: Developing farmer participatory research methods in Mexico. *Development in Practice, 11*(4), 417–424.

Enxian, H. (2002). Investigation report on forestry in The DPRK [J]. *Shaanxi Forestry Science and Technology, 1*, 90–92 (in Chinese).

Hayes, P. (2009). Unbearable legacies: The politics of environmental degradation in The DPRK. *Asia Pacific Journal., 41*, 1–9.

He, J., Ho, H., & Xu, J. (2015). Participatory tree species selection for sloping land management in North Korea. *Mountain Research and Development, 35*(4), 318–327.

Hu, J., & Xu, J. (2017). Is there decentralization in North Korea? Evidence and lessons from the Sloping land management program 2004–2014. *Land Use Policy, 61*, 113–125.

Hippel, D. V., & Peter, H. (2007). Energy security for The DPRK. *Science, 316*, 1288–1289.

Hobbs, J. R., Higgs, E., Hall, M. C., Bridgewater, P., Chapin, S. F., III., Ellis, C. E., Ewel, J. J., Ewel, J. J., Hallett, L. M., Harris, J., Hulvey, K. B., Jackson, S. T., Kennedy, P. L., Kueffer, C., Lach, L., Lantz, T. C., Lugo, A. E., Mascaro, J., Murphy, S. D., Nelson, C. R., ... Yung, L. (2014). Managing the whole landscape: Historical, hybrid, and novel ecosystems. *Frontiers Ecology Environment, 12*, 557–564.

Hong, S. J. (1999). Environmental pollution in The DPRK: Another South Korean burden? *East Asian Review, 11*, 79–98.

Huizhi, Z. (2010). Adjustment of The DPRK's economic development strategy in the new era. *Northeast Asia Forum, 3*, 17–24 (in Chinese).

IISD. (2002, February 27–28). Summary of the international expert meeting on forest landscape restoration. *Sustainable Developments, 71*(1). (2 March 2002). http://www.iisd.ca/crs/sdcfr/sdvol71num1.html. Accessed 3 March 2015.

International Food Policy Research Institute (IFPRI). (2015). Global hunger Index—Armed conflict and the challenge of hunger.

Jin, Z., Huizhi, Z., & Mingxu, C. (2016). Environmental management and sustainable development in Korea. *Population Journal, 04*, 88–92 (in Chinese).

Jozuka, E. (2016, September, 30). The DPRK floods: Survivors homeless, sick as winter looms. *CNN*. https://edition.cnn.com

Kang, S., & Choi, W. (2014). Forest cover changes in The DPRK since the 1980s. *Regional Environmental Change, 14*(1), 347–354.

Katila, P., Colfer, C. J. P., de Jong, W., Galloway, G., Pacheco, P., Winkel, G. (Eds.). (2019). *Sustainable development goals: Their impacts on forests and people*. Cambridge University Press.

Kim, H. B, & Jang, D. S. (1994). *Maulsoop, the Korean village grove* (p. 204). Yeolhwadang Publishing Co., (in Korean).

Kim, I. A, Youn, Y. C. (2005, August 8–13). Conditions for sustainable community forests: Evidence from 17 village groves. In Korea. In J. I. Innes, I. K. Edwards, & D. J. Wilford (Eds.), *Forests in the balance: Linking tradition and technology*. Abstracts of the XXII IUFRO World Congress, *International Forestry Review, 7*(5), 221.

Kim, J. H., et al. (2010). Biodiversity in the context of management and conservation of forest resource. *Journal of Forest Science, 26*(2), 65–73.

Kim, R. E. (2015). *Legal brief on the Baekdudaegan protection act, 2003*. International Development Law Organization.

Kim, R. E., & Ali, S. H. (2016). Green diplomacy—An opportunity for peace-building? *Environmental Policy and Law, 46*(1), 86–96.

Kim, S. H., et al. (2011). The Northeast Asian energy situation and the The DPRKn factor. *The DPRKn Review, 7*(1), 80.

Kumtashula, E., & Mafongoya, P. L. (2005). Farmer participatory evaluation of agroforestry trees in eastern Zambia. *Agricultural Systems, 84*, 39–53.

Kyung-Seok, P., Lee, S.-Y., & Park, S.-Y. (2013). Changes in The DPRK's forestry policy. *The DPRKn Review, 9*(1), 20–29.

Lamb, D., Erskine, P. D., & Parrotta, J. A. (2005). Restoration of degraded tropical forest landscapes. *Science, 310*(5754), 1628–1632.

Laurance, W. F., Sayer, J., & Cassman, K. G. (2014). Agricultural expansion and its impacts on tropical nature. *Trends in Ecology & Evolution, 29*, 107–116.

Lee, D. (2008, October 5–10). A Korean perspective of traditional knowledge forest-related knowledge. In *Proceedings of the 1st international conference on forest related traditional knowledge and culture in Asia* (pp. 3–5). Korea Forest Research Institute, Korea Forest Research Institute.

Lee, G. (2003). The political philosophy of Juche. *Stanford Journal of East Asian Affairs, 3*, 105–112.

Li Shixian. (1985). Management of Forest Industry in Korea. *Forestry Economy, 1*, 61–63 (in Chinese).

Li., Xueqiang, & Yutao, G. (2007). The DPRKn agriculture in Kim Jong Il's Era: Seeking change in difficulties. *Research World, 6*, 37–39 (in Chinese).

Liu, J. (2003). Support to private and community farm forestry in China. *Unasylva 212, 54* (1), 57–62.

Liu, J. (2006). *Forests in the mist—Livelihoods and responses to the natural forest protection program in China* (Wageningen PhD Dissertation, p. 87).

Liu, J. (2009). Reconstructing the history of forestry in Northwestern China 1949–1998. *Global Environment, 4*, 188–219.

Liu, J., Liang, M., Li, L., Long, H., & De Jong, W. (2016). Comparative study of the forest transition pathways of nine Asia-Pacific countries. *Forest Policy and Economics, 76*, 25–34. https://doi.org/10.1016/j.forpol.2016.03.007

Liu, J., & Innes, J. (2015). Participatory forest management in China: Key challenges and ways forward. *International Forestry Review, 17*(4), 477–484.

Liu, L., Liu, M., Xu, Y., & Liu, J. (2015). Changes and tendency analysis of forest land area in the key stated-owned forest region: A case study based on the data of forest industry region of Heilongjiang province. *Scientia Silvae Sinicae, 52*(8), 88–94.

Liu, Z., & Li, S. (1984). Investigation report on forestry management in The DPRK. *Agricultural Economic Issues, 7*, 61–64 (in Chinese).

Long, H., Liu, J., Tu, C., Fu, Y. (2018). From state-controlled to polycentric governance in forest landscape restoration: The case of the ecological forest purchase program in Yong'an municipality of China. *Environmental Management, 62*(1), 58–69.

Luo, Y., Liu, J., Zhang, D., & Dong, J. (2015). Actor, customary regulation and case study of collective forest tenure reform intervention in China. *Small-Scale Forestry, 14*(2), 155–169. https://doi.org/10.1007/s11842-014-9279-1

Mansourian, S., Vallauri, D., & Dudley, N. (2005). *Forest restoration in landscapes: Beyond planting trees*. Springer.

Mansourian, S., & Vallauri, D. (2014). Restoring forest landscapes: Important lessons learnt. *Environmental Management, 53*(2), 241–251.

MoLEP (Ministry of Land and Environmental Protection). (2010a). *Report of national forest resources inventory*.

MoLEP. (2010b). *Gereral forest resources inventory*.

MoLEP. (2011). *DPRK 4th report on CBD*.

MoLEP. (2012). *Democratic people's republic of Korea environment and climate change outlook, ministry of law and and environment protection*, DPRK, (ISBN 978-9946-1-0170-5).

MoLEP. (2015). *DPRK national Agroforestry strategy and action plan*.

Monroe, A. S., & Butler, W. H. (2016). Responding to a policy mandate to collaborate: Structuring collaboration in the collaborative forest landscape restoration program. *Journal of Environmental Planning and Management, 59*(6), 1–19.

Na, H. R. (2012). Nationalism as a factor for an international environmental regime: Korea and the East Asian Biosphere Reserve Network (EABRN). *East Asian Science, Technology and Society, 6*, 83–99.

Nam, S. (2004). The legal development of the environmental policy in the democratic people's republic of Korea. *Fordham International Law Journal, 27*(4), 1322–1342.

Newton, A. C., Del Castillo, R. F., Echeverría, C., Geneletti, D., González-Espinosa, M., Malizia, L. R., Premoli, A. C., Benayas, J. M. R., Smith-Ramirez, C., & Williams-Linera, G. (2012). Forest landscape restoration in the drylands of Latin America. *Ecology and Society, 17*(1). https://doi.org/10.5751/ES-04572-170121

Noland, M. (2004). Famine and reform in The DPRK. *Asian Economic Papers, 3*(2), 1–40.

Ostrom, E. (2000). Beyond markets and states: Polycentric governance of complex economic systems. *The American Economic Review*, 641–672.

Ostrom, E. (2010). Polycentric system for coping with collective action and global environmental change. *Global Environment Change, 20*, 550–557.

Pang, C., Yu, H., He, J., & Xu, J. (2013). Deforestation and changes in landscape patterns from 1979 to 2006 in Suan County, DPRK. *Forest, 4*, 968–983.

Park, J. H. (2010). A research on the actual condition of The DPRK's forest devastation through MODIS satellite images. In *The North and South' s cooperation for The DPRK's forestation* (p. 7). Institute for Peace and Unification Studies at Seoul National University.

Park, K. S., Lee, S. Y., & Park, S. Y. (2013). Changes in North Korea's forest policy. *North Korean Review, 9*(1), 20–29.

Park, P. H. (ed.). (2009). *The dynamics of change in The DPRK: An institutionalist perspective*. Institute for Far Eastern Studies, Kyungnam University. Boulder.

Park Kyung-Seok et al. (2008, December 8). *A study on the trait of the forest system of the central and local governments and its management system in The DPRK*. Korea Forest Research Institute, Research Report.

Parrotta, J. A., Lim, H. F., Liu, J., Ramakrishnan, P. S., & Youn, Y.-C. (2009). Traditional forest-related knowledge and sustainable forest management in Asia. *Forest Ecology and Management, 257*, 1987–1988.

Reed, J., Van Vianen, J., Deakin, E. L., Barlow, J., & Sunderland, T. (2015). Integrated landscape approaches to managing social and environmental issues in the tropics: Learning from the past to guide the future. *Global Change Biology, 22*, 2540–2554.

Reubens, B., Moeremans, C., Poesen, J., Nyssen, J., Tewoldeberhan, S., Franzel, S., Deckers, J., Orwa, C., & Muys, B. (2011). Tree species selection for land rehabilitation in Ethiopia: From fragmented knowledge to an integrated multi-criteria decision approach. *Agroforestry Systems, 82*(3), 303–330.

Ribot, J. C. (2005). Choosing representation: Institutions and powers for decentralized natural resources management. In J. Carol, P. Colfer & D. Capistrano (Eds.), *The politics of decentralization* (pp. 86–106). Earthscan.

Ros-Tonen, M. A. F., Derkyi, M., & Insaidoo, T. F. G. (2014). From co-management to landscape governance: Whither Ghana's modified taungya system? *Forests, 5*(12), 2996–3021.

Sabogal, C., Besacier, C., & McGuire, D. (2015). Forest and landscape restoration: Concepts, approaches and challenges for implementation. *Unasylva, 245, 66*(3), 3–10.

Sayer, J., Chokkalingam, U., & Poulsen, J. (2004). The restoration of forest biodiversity and ecological values. *Forest Ecology Management, 201*, 3–11.

Sayer, J., Sunderland, T., Ghazoul, J., Pfund, J. L., Sheil, D., Meijaard, E., ... & Buck, L. E. (2013). Ten principles for a landscape approach to reconciling agriculture, conservation, and other competing land uses. *Proceedings of the National Academy of Sciences of the United States of America, 110*, 8349–8356.

Sayer, J., Margules, C., Boedhihartono, A. K., Dale, A., Sunderland, T., Supriatna, J., & Saryanthi, R. (2014). Landscape approache: What are the pre-conditions for success? *Sustainability Science, 10*, 345–355.

Seymour, F., & Busch, J. (2017). *Why forests? Why now? The science, economics and politics of tropical forests and climate change*. Center for Global Development.

Scherr, S., Shames, S. A., Friedman, R. (2013). *Defining integrated landscape management for policy makers* (Ecoagriculture Policy Focus. No. 10). Ecoagriculture Partners.

Sharma, N. P. (1992). A global perspective on forest policy. In N. P. Sharma (Ed.), *Managing the world forest: Looking for balance between conservation and development* (pp. 17–33). Kendal/Hunt Pub. Co.

SOFO. (2018). *State of the World's forest 2018*. Food and Agricultural Organiztion.

Song, A. Y., & Hastings, J. V. (2020). Engaging The DPRK: Environmental cooperation in peace building. *Third World Quarterly, 41*(11), 1809–1927.

Stanturf, J., Lamb, D., & Madsen, P. (2013). *Forest landscape restoration[M]*. Springer.

Tang, L., Shao, G., Piao, Z., Dai, L., Jenkins, M. A., Wang, S., Wu, G., Wu, J., & Zhao, J. (2010). Forest degradation deepens around and within protected areas in east Asia. *Biology Conservervation, 143*(5), 1295–1298.

Tian Gang. (2005). *Research on regional forestry economic and trade cooperation in Northeast Asia [D]*. Northeast Forestry University (in Chinese)

Tscharntke, T., Tylianakis, J. M., Rand, T. A., Didham, R. K., Fahrig, L., Batáry, P., ... & Westphal, C. (2012). Landscape moderation of biodiversity patterns and processes—Eight hypotheses. *Biological Reviews, 87*, 661–685.

Tu, C., Hexing, L., & Jinlong, L. (2016). Experience of soil and water loss control in Changting, Fujian Province from the perspective of forest landscape restoration. *Forestry Economy, 03*, 14–18 (in Chinese).

UNDP. (2011). *Country programme for the democratic people's Republic of Korea (2011–2015)*.

UNDP. (2018). *DPRK needs and priorities*. https://www.undp.org/content/dam/unct/dprk/docs/unct_kp_NP2018.pdf

UN Environmental Assembly. (2019). *National statement of democratic people's republic of Korea at the fourth session of the United Nations environment assembly*. http://web.unep.org/environmentassembly/democratic-peoples-republic-korea

United Nations Population Fund (UNFPA). (2014). DPRK socio-economic, demographic and health survey; DPRK Central Bureau of Statistics. (2008). DPRK 2008 population census—National report.

Unite Nations Population Fund (UNFPA). (2008). *DPRK population census national report*.

Van Oosten, C. V. (2013). Restoring landscapes—Governing place: A learning approach to forest landscape restoration. *Journal of Sustainable Forestry, 32*(7), 659–676.

Van Oosten, C., Gunarso, P., Koesoetjahjo, I., & Wiersum, F. (2014). Governing forest landscape restoration: Cases from Indonesia. *Forests, 5*, 1143–1162.

Van Oosten, C., Runhaar, H., & Arts, B. (2021). Capable to govern landscape restoration? Exploring landscape governance capabilities, based on literature and stakeholder perceptions. *Land Use Policy*. https://doi.org/10.1016/j.landusepol.2019.05.039

Wang, E., Guang, L., & Tian, G. (1992). Current situation and prospect of regional forestry economic cooperation in Northeast Asia. *Forestry Economy, 5*, 42–46 (in Chinese).

West, P. C., Gerber, J. S., Engstrom, P. M., Mueller, N. D., Brauman, K. A., Carlson, K. M., Cassidy, E. S., Johnston, M., MacDonald, G. K., Ray, D. K., & Siebert, S. (2014). Leverage points for improving global food security and the environment. *Science, 345*, 325–328.

Wiersum, K. F., & Ros-tonen, M. (2005). *The role of forests in poverty alleviation: Dealing multiple millennium development goals*. Wageningen University.

Winstanley-Chesters, R. (2015). *Environment, politics, and ideology in The DPRK: Landscape as a political project*. Rowman & Littlefield.

World Bank. (2001, July 30). *A revised forest strategy for the World Bank Group*. (Draft). The World Bank. www.worldbank.org.

Xiaodan, S. (2009). The DPRK's food problem and international aid channel. *Foreign Economic and Trade Practice, 1*, 35–37 (in Chinese).

Xu, J., Noordivijk, M., He, J., Kim, K.-J., Jo, R.-S., Pak, K.-G., Kye, U.-H., Kim, J.-S., Kim, K.-M., Sim, Y.-N., Pak, J., Song, K.-U., Jong, Y.-S., Kim, K., Pang, C.-J., & Ho, M. (2012). Participatory agroforestry development for restoring degraded sloping land in DPRK. *Agroforestry Systems, 85*(2), 291–303.

Xu., & Guoyi. (1994). Significant changes in forest ownership in Korea. *World Forestry Research, 4*, 61 (in Chinese).

Xu, J. (2012). China's new forests aren't as green as they seem. *Nature, 477*, 371.

Youn, Y.-C. (2009). Use of forest resources, traditional forest-related knowledge and livelihood of forest dependent communities: Cases in South Korea. *Forest Ecology and Management, 257*, 2027–2034.

Youn, Y.-C., Liu, J., Daisuke, S., Kim, K., Ichikawa, M., Shin, J.-H., & Yuan, J. (2011). Northeast Asia (Chapter 8). In J. A. Parrotta & R. L. Trosper. (Eds.), *Traditional forest-related knowledge: Sustaining communities, ecosystems and biocultural diversity* (pp. 281–313, Vol. 12). World Forest Series, Springer.

Yutao, G. (2007). The DPRK's economic reform and its characteristics shrouded in mystery and particularity. *Research World, 2*, 39–40. (in Chinese).

Zhang Xiaohong. (2010). *Research on forest landscape restoration [D]*. Chinese Academy of Forestry Sciences (in Chinese)

Zhang, S., Zhu, C., & Xiao, W. (2000). *Introduction of sustainable forestry management* (pp. 61–63). China's Forestry Publishing House (in Chinese).
Zhengji, C. (2014). Reclamation of Xipu Hills in The DPRK. *Northeast Asia Forum, 1*, 64–65 (in Chinese).

Index

L. Jinlong, *Crying Forests*,
https://doi.org/10.1007/978-981-16-9725-8

D

E

F

G

The manufacturer's authorised representative in the EU is Springer Nature Customer Service Centre GmbH, Europaplatz 3, 69115 Heidelberg, Germany. If you have any concerns regarding our products, please contact ProductSafety@springernature.com

Printed and bound by CPI Group (UK) Ltd, Croydon, CR0 4YY
30/09/2026
02225235-0001